KV-681-251

Donough O'Brien spent many years in PR and marketing before retiring to write. This is his fourth book (*Fringe Benefits, Fame by Chance, & Banana Skins*). Donough specialises in collecting amazing historical facts.

Anthony Weldon is a non-fiction publisher and has published Donough's three previous books. His company, Bene Factum, specialises in publishing biographies, general information, travel and business books, as well as managing bespoke publishing projects.

Any number of things you didn't know . . .

and some you did

Donough O'Brien & Anthony Weldon

Bene Factum Publishing

First published in 2008 by
Bene Factum Publishing Ltd
PO Box 58122
London SW8 5WZ

Email: inquiries@bene-factum.co.uk
www.bene-factum.co.uk

ISBN: 978-1-903071-18-2

2 4 6 8 10 9 7 5 3 1

A CIP catalogue record of this is available from the British Library

Cover design by Paul Fielding Design, Putney, London
Cover illustration © The Saul Steinberg Foundation/ARS, NY and DACS London 2008
Typeset by Tony Hannaford of 01:01 Design Consultants, Putney, London
Numbers graphic designed by Conor Bevan, London
Website designed and developed by Conor Bevan and James Wilson Jennings

Printed and bound by Clays Ltd, St Ives Plc, Bungay, Suffolk

Contents

Introduction

Buried somewhere in this book are two lamentably low numbers relating to the authors' youthful lack of mathematical competence. So why did we create 'Numeroids' - a book about numbers?

Because numbers go so much further than maths. Quite simply, they govern our world – both overtly and covertly. Every day we use PIN numbers, work out the cost of things, rely on numbers to plan and organise, use numbers in figures of speech, study sport scores, watch our speed limits, check our bank balances, hope our lucky numbers will come up in the lottery, and so on...and on. And that's just the start.

On top of that, there's a vast legion of hidden numbers, such as those that drive our computers, make up the constituents of our food, calculate risks, regulate our bodies, run the economy and keep the natural world ticking over.

We have selected our 'Numeroids' completely at random, with the sole purpose of engaging and fascinating our readers. There's something here for everybody. We've had good fun collecting and explaining them, and hope that discovering them will be equally enjoyable.

Considerable research has gone into our attempts to ensure that all of our 'Numeroids' are correct (in many cases more challenging than we could have ever imagined) and many hours have been spent cross-checking numbers in the public domain. Our sources have been wide and varied. 'Numeroids' is not intended as a reference book, rather as a book to entertain. However, if you think any of our facts are wrong, please let us know.

And please give us more 'Numeroids' for our next edition!

A number of thanks are due, especially to Liz Cowley for her indefatigable proof-reading – we are in awe of her eagle eye; to Paul Fielding for unearthing the Saul Steinberg cartoon for his stunning cover design; to Tony Hannaford and his colleague Kate Cooper, who conquered the challenging job of typesetting over 1,200 numbers in sections and ascending order; to Conor Bevan and James Wilson Jennings for their tremendous 'Numeroids' website; and, finally, to all those who have leaped on the 'Numeroids' bandwagon, shared our enthusiasm and made so many great suggestions for entries. We are indebted to you all.

Donough O'Brien and Anthony Weldon

www.numeroids.com

Numeroid Numbers

7.8 x 5

Numeroids' page size is 'Large Crown Octavo' – 7.8 x 5 inches (198 x 129 mm). This format, also known as 'trade paperback' or 'B format', is based on the old Crown paper size of 15 x 20 inches (380 x 508 mm). The 'Octavo' refers to the fact that the sheet was originally folded three times to make eight sides of paper.

49.5

In terms of UK book sales, bookshops account for 49.5%, the internet 16.5%, supermarkets 9.8% and other outlets 24.2%.*

1,321

Individual entries in Numeroids.

5,310

The number of retail book shops in the UK, of which 1,424 are independents and not part of multiple chains. **

56,050

Numeroids has 56,050 words in the whole book.

118,602

The amazing number of new book titles published in 2007 in the UK (up 36% from the previous year). The number of available backlist titles (i.e. books published pre-2007) is 758,125 – an indication of the crowded market in which this book has to compete.

257,404

The worldwide number of new books published in the English language in 2007. *

* Source: Booksellers' Association ** Source: Publishers' Association

www.numeroids.com

Modern Times

1

Just under one acre (0.4 of a hectare) is the average space per person in the UK – with a population of just over 60 million people and a landmass of 59,754,931 acres (24,182,520 ha) - slightly smaller than the US State of Oregon. In Ireland, with a population of 4.1 million and an area of 16,937,480 acres (6,854,194 ha), there are 4.1 acres (1.7 ha) per inhabitant.

2

Most common number used in mobile phone texting, standing for 'to' or 'too', closely followed by 4 in place of 'for'.

2

Number of weeks an average person spends kissing in a lifetime.

2

The amount of technical information available is doubling every two years, but by the year 2010 it is estimated that this could double every 72 hours.

3

About one third of women may look as if they are blonde, but in fact only 3% are naturally so!

3

The average human spends 3 years of life on the WC.

3

Three piece suite; when the 'suite' is spelt with an 'e', it refers to the average requirement of furniture for a sitting room – a sofa and two armchairs. However, if the 'suit' has no 'e' then this refers to clothing – normally a man's suit with jacket, trousers and waistcoat made of the same cloth.

4
Four is the unluckiest number in China, probably because it sounds like 'death' in Cantonese. So strongly is this felt that, in many places, there are no 4th floors in buildings and no 4 in car number plates.

4
Victoria Beckham has 4 eight-pointed stars tattooed at the base of her spine – one each for herself, David, Brooklyn and Romeo. At the last count, husband David had 11 tattoos.

4.5
The fastest population growth in any country is in Liberia with 4.5% a year. By contrast, in India it is 1.46%, Ireland 1.33%, USA 0.97%, China 0.58% and the UK 0.42%.

6
For London black cab drivers to pass their 'Knowledge' test, they must know all the streets within a six-mile radius of Charing Cross Station, as well as 320 basic routes and all the main places of interest; theatres, etc. One of the authors was assured by one particular cabbie that "he knew well over 70,000 facts when he first became a driver"!

6.6
The successful romantic novel publisher Mills & Boon sells a paperback every 6.6 seconds in the UK. Its worldwide market accounts for 200 million books a year.

7
7 litres of water and 162 grams of oil are required to create a one-litre plastic water bottle. In 2006, 2.27 billion litres of bottled water were drunk in the UK.

7
Renewable sources of energy – solar, wind and tidal – already supply 7% of the world's energy.

8
Eight hours of video material is uploaded every minute on to the You Tube website.

10

Almost ten per cent of the world's above ground stock of gold is held in India in the form of rings, necklaces and jewellery – more than 14,000 tons – and considered a good investment. In contrast, less than 2% of the Indian population invests in stocks and shares.

10

Percentage of attempted suicides that are successful.

11

The duration of the average marriage in the UK is about 11 years. This has dropped dramatically from the 1980 average of 37 years. The rate is 14 divorces per 1,000 of the married population, with the highest percentage in men and women aged 25 to 29. However, marriage still remains a popular institution, with 91% of men and 95% of women being married by the age of 50. The age of marriage has risen from 25.6 for men and 23.1 for women in 1961 to over 30 and 28 respectively.

13.4

Ireland's per capita alcohol consumption increased from 7.0 litres in 1970 to 13.4 by 2006. This increase resulted in 1,755 deaths. This compares with the French 1970 consumption of 20.4 litres, dropping to 13 today.

15

15% of the world's tower cranes are based in Dubai to support its phenomenal building boom.

15

The average number of positive responses a spammer receives from sending out 1 million spams. The first spam is thought to have been sent out in 1978 to the 398 users of the US Government's Aparnet – the forerunner of the internet. It is estimated that only 200 spammers are responsible for 80% of the world's spams, which in the USA alone extracted $240 million from recipients. Fighting these illegal emails is even bigger business, thought to be worth $140 billion.

15

The percentage of non-smoking girls who admitted they had had sex at American universities in the 1970s. However, of girls who smoked, 55% had had sex. A helpful pointer for the observant male student?

16

Years the Starbucks logo showed a Norse siren before it was altered to cover her bare breasts.

20

Britons spend 20% of their income on leisure pursuits. In the mid-1950s, this figure was only 9%. In the UK, about 14% of household income is now spent on food, 9% on fuel, 3% on education, and 3% on health. In comparison, an Irish household spends 21% of its income on food, 10% on fuel, 7% on education and 4% on health.

20

The UK has 20% or 1/5th of all the world's CCTV cameras – one for every 14 of us.

21

By the time they have reached the age of 21, most Americans have spent 20,000 hours (2.3 years) watching TV, 10,000 hours playing video games and the same amount of time on the phone.

33

Thirty-three per cent of the world's population eat with a knife and fork, whilst the same percentage eat with chopsticks.

40

Gallons – the average use per head of domestic water a day in Britain (USA is 80).

43

Square feet - the size of Britain,s smallest office, the so-called 'Tardis' in Truro, Cornwall - 7 feet x 6 feet, (2 x 1.8 m). In 2008, it was on the market for £19,950 on 125 year lease.

47

Forty seven KitKat chocolate bars are eaten every second in the UK – the country's favourite chocolate bar.

50

Per cent of all 16-19 year old boys in Britain have never written a letter, not even to thank for a present!

66

When most of the safety features were removed from London's Kensington High Street, the incidences of pedestrians and bicyclists 'Killed or seriously injured' (KSIs) actually fell by 60%. Why? Because with so many of these features in place, it is thought that they took away responsibility from both drivers and pedestrians, thus making accidents more likely.

70

Seventy per cent of American four year olds have used a computer. 52% of Korean infants between the age of 3 and 5 use the internet, and they spend an average of 4 hours a week online.

75

Amongst British Christians, almost 75% believe that God is male.

80

The legal 'drink drive' limit in the UK is 80 mg of alcohol per 100 millilitres of blood – it is difficult to translate this into specific units of alcohol as its effect relates to quite a number of factors – food eaten, and size and sex of the drinker. However, 34% of men and 23% of women still drink and drive.

92

The population of Ireland is predominantly Catholic – 92%. However, it is thought that only about 50% are active worshippers. In the UK, the Catholic congregation is about 9 million, recently boosted considerably by the Polish immigrant community.

94.5

Per cent of the prison population in the UK which is male, with only 5.6% female, of which 40% have been convicted of drug offences. Only 16% of male prisoners are drug offenders.

110

Britons use an average of 110 rolls of toilet paper per head each year – considerably more than the EU average of 77.5. Consumption in Western Europe and the US averages out at 8.6 sheets per visit.

130

It takes, on average, 130 telephone calls to ensure that an A-list celebrity appears at an award ceremony such as the BAFTAs.

146

Every minute the world's population increases by 146, with 249 births set against 103 deaths.

240

The population of the Falkland Islands is 3,105, but humans are outnumbered by sheep by a factor of 240 to 1, and by 330 to 1 by penguins.

554

If you dropped $554 on the ground, it just wouldn't be worth Microsoft founder Bill Gates's time to bend down and pick them up if it took him longer than 5 seconds. His estimated income is $6,659 per minute (or $110.90 per second). However, his charitable foundation does give away about $2,663 per minute.

1,000

The number of megabytes in a gigabyte, but with the increased capacity in computing power, this is fast becoming obsolete as a unit of storage. The terabyte (1,000 gigabytes or 1,000,000 megabytes) will soon be the necessary norm on most computers. On the horizon is the even larger unit, the petabyte (1,000 terabytes), and in its wake there are the exabyte, the zettabyte, the yottabyte and ultimately the brontobyte (1,000,000 petabytes) and ...well ...if you are still reading this ...we haven't really got room to write down how many gigabytes this is.

1,080

The latest full High Definition (HD) TV screen has 1,080 vertical lines of pixels on a screen. Regular HD has 720, while the older regular definition has only 576.

2,752

In the 9/11 aircraft attacks on the World Trade Center in New York in 2001, 2,752 people died as an immediate result, including members of at least 80 different nations. Another 24 were reported 'missing, presumed dead'. There have been subsequent deaths from lung disease as a result of the massive quantities of dust caused by the explosions. These figures exclude the 10 hijackers.

3,176

With about 34 million vehicles on the roads, there are a relatively low proportion of deaths – 3,176. This is especially so when compared with the much higher number of 7,000 deaths in the early 1930's with only 2.4 million vehicles. The first national speed limit introduced in 1930 and then the compulsory use of seatbelts were two factors that have made all the difference.

6,726

Boxes of cornflakes produced from 1 acre of harvested corn.

8,333

There are many factors that affect the amount of paper produced from a tree – such as the size and type of wood and method of pulping. However, on average, 100% virgin pulp from 12 trees produces 1 ton of newsprint, which in turn produces 14,000 average tabloid newspapers. Or, put another way, 1 tree can produce 8,333 sheets of copier paper.

8,500

Pounds – the amount an average woman will spend on make-up in her lifetime. Foundation is the highest expense at £1,998, followed by lipstick at £1,342, mascara at £1,404, and eyeshadow at £1,144. 90% of women would not go out without make-up, feeling they are 'naked'.

10,000

Children can be exposed to as many as 10,000 TV ads a year.

13,080

Of the 40,000 people reported 'missing, presumed dead' as a result of the conflicts in the former Yugoslavia, over 13,000 have been identified by the International Commission for Missing

Persons, using 85,640 blood samples and 248,586 bone samples. The Commission has carried out similar work for the 2004 Asian tsunami and the 2005 Hurricane Katrina disaster in Louisiana, USA.

13,087

There are over 13,000 licenced yellow taxis in New York City, with 42,000 drivers, and they carry 241 million passengers a year. The number of licences or 'medallions' is limited, and they are so expensive, at $200,000 each, that they are often owned by finance houses. In 2006, Gene Friedman bought 84 as an investment for $25 million, or $477,000 (£238,000) each. In comparison, a London taxi licence costs only £178.

20,000

The horrific number of the 7 million commuters on the Indian railways in Mumbai (Bombay) who have been killed in the last five years. This is because many of them have to sit on the roof or hang off the sides of the carriages, and are swept off or electrocuted.

26,831

There are 26,831 health clubs in the USA. In 1991, there were only 200 registered Pilates instructors in America – now there are 14,000.

37,000

Some British women spend a total of about £37,000 on their hair in a lifetime, and a total of two years tending to their hair – 41 minutes a day.

48,328

The number of churches in the UK includes 18,503 Anglican, 4,585 Catholic and 6,062 Methodist. These are served by 35,289 Ministers of whom 10,041 are Anglican, 6,239 Catholic and 2,464 Presbyterian.

68,000

In 1997, 68,000 UK residents claimed non-domestic status – but by 2006 this figure had risen to at least 114,000.

210,000

There are 210,000 people on the reported Missing Persons List in the UK, of whom 140,000 are under 18.

270,000

The number of cars stolen each year in Britain. But, thanks largely to better anti-theft devices, this is 61% less than when car theft was at its peak in 1995.

1,000,000

Gallons – the amount of urine spread by dogs on London's parks each year.

1,240,000

In the UK there are over a million people between the ages of 15 and 24 who are classified by the acronym NEET (Not currently engaged in Education, Employment or Training).

2,000,000

Every Muslim is obliged to travel to Mecca at least once in a lifetime. The Hajj is the world's largest pilgrimage, with two million pilgrims each year.

4,000,000

Ilja Gort, a Dutch wine producer, has insured what he considers to be his most important business asset – his nose, with its unusually advanced sense of smell – on which he relies to produce top quality wine. The risk is for £4 million, placed at Lloyd's of London. It is not unusual to insure vital body parts at Lloyd's, and other unusual policies include food critic Egon Ronay's tastebuds (£250,000), Bruce Springsteen's voice (£3.5 million), Betty Grable's legs ($1 million) and Dolly Parton's breasts (£350,000 – the pair).

4,000,000

The profile of church attendance and religious worship in the UK is in a state of change. Currently there are about 4 million regular Christian churchgoers. But trends indicate that by the year 2050 this may have fallen to just under 1 million. However, Hindu worshippers will have expanded from 400,000 to 855,000, and by 2050 there will be over 2.5 million active Muslims.

5,900,000

Using the definition of consuming more than 8 units per day of alcohol for men and 6 for women, then an amazing 5.9 million people 'binge' drink. As a consequence, it is estimated that drink-related crime and anti-social behaviour costs the country £7 billion per year.

6,000,000

It is estimated that six million people have some kind of accident – small or large – in the street while texting on their mobile phones. By contrast, only 1,400 have accidents with steam irons. A sign of the times?

24,300,000

Sir Paul McCartney paid Heather Mills £24 million in their divorce settlement after four years of marriage, although she had asked for at least three times as much. However, the largest British divorce payout was for £48 million in 2006 by John Charman, an insurance magnate. These were dwarfed in America by Michael Jordan and Neil Diamond who each had to pay the equivalent of £75 million to their respective ex spouses.

25,000,000

According to the RAC, of the 25 million people in Britain who commute, 71% travel to work by car, 11% walk, 8% by bus, 6% by rail, 3% cycle and 1% by motorbike. Britons have the longest journeys to work of any of the main European nations – an average trip of 54 minutes a day – and 3% travel an excess of 3 hours a day going to and from work.

27,000,000

About 27 million eggs are eaten daily by Britons (about 10 billion every year). Forty per cent are classified as 'free range'.

47,000,000

The 31,000 McDonalds restaurants worldwide serve over 47 million customers every day in 119 countries, employing 1.5 million people.

100,000,000

Estimated number of acts of sexual intercourse that take place worldwide each day.

335,000,000

In Britain, 335 million tons of solid waste are thrown away every year, including 12 billion aluminium cans and 6 billion bottles. Only 17% is recycled.

1,200,000,000

1.2 billion SMS text messages are sent by mobile phone every week in the UK.

1,244,000,000

The number of the world's population that has access to the internet. 31% of them use the English language, followed by 16% Chinese, 9% Spanish, 7% Japanese and 5% French. In proportion, 37% of the internet's use is in Asia, 27% in Europe and 19% in North America.

1,695,820,000

Over 1.6 billion is the projected population of India by 2050, overtaking China with 'only' 1.47 billion.

2,600,000,000

The enormous number of the world's population lacking sanitary toilet facilities.

9,000,000,000

It is estimated that by the year 2042, the world's population will have grown from its current 6.7 billion people to 9 billion. The current rate of increase is 1.14% (about 211,090 per day) but this is down from the peak rate of 2.19% in 1963. In the forty years between 1959 and 1999, the world's population doubled from 3 billion to 6 billion. By the year 2000, there were 10 times as many people as there were 300 years ago.

25,000,000,000

The Japanese get through 25 billion pairs of wooden chopsticks each year. However, the Chinese use 45 billion disposable chopsticks – consuming millions of birch, poplar, and bamboo trees a year.

-3

It was 3 months before he was born when Shapur II became King of Persia. After his father died, he was crowned in 309 when the ruling nobles reserved the throne for the as yet unborn child by placing the crown on the stomach of his six months pregnant mother.

0

'Nothing', wrote Louis XVI in his diary on 14 July 1789, the day the Paris mob stormed the Bastille, an act that sealed his fate and ultimate execution.

0

Although Berengaria (c1165-1230) was Queen of England as the wife of King Richard I (the Lionheart) and his Queen Consort, she spent no time in England. Born in Navarre, she married Richard in 1191 in Limassol, Cyprus, while he was on his Third Crusade. When Richard was imprisoned, she stayed in Europe trying to raise his ransom, and on his liberation he only spent three months in England in the course of their marriage, during which time he was forced to concentrate on retrieving his kingdom from his brother John.

0

Ambassadors that the United States had in place in foreign countries before 1893. As a republic, the US had Ministers, but Ambassadors were considered a throw-back to royalty.

0

Aristocratic women in Venice during the Renaissance who were not blonde – fashion dictated that they dyed their normally dark hair.

0

Despite the fact that the name 'Napoleon Brandy' is widely recognised, the Emperor drank virtually no brandy. Although he was fairly abstemious, his preferred drink was a red Burgundy – Chambertin. However, Napoleon did take some barrels of

Courvoisier brandy to exile in St Helena, which was much appreciated by the accompanying English officers, and it is they who called it 'The Brandy of Napoleon'.

0

Legitimate children sired by the 'Merry Monarch', King Charles II. However, he produced a number of illegitimate ones – as many as fourteen have been recognised from at least seven different mistresses. The Duke of Buckingham referred to Charles as "the father of his people", adding "...a good many of them".

0

The amount of dealings or communications with Captain Charles Boycott – Lord Erne's agent in County Mayo, when in 1880 he was ostracised by the local community for undermining the Land League's attempts to protect the tenants. This gave the English language a new word – 'boycott'.

0

The first recorded use of a symbol for zero was in the 3rd century in Babylon, where inhabitants solved the problem of representing 'nothing' in a written numeral by giving it a symbol. Independently, the Mayan civilisation came up with an answer to the same problem in the 3rd century AD. Hindus in India also developed their own symbols, and the first recorded is thought to be from Gwalior.

0

Years – the amount of time that Elihu Yale lived in America after he was 3. Back in England, he sent some small gifts to the Collegiate School of Connecticut, because of which it rather generously changed its name to Yale University.

0.5

When America banned alcohol during Prohibition in 1920, there was in fact a permitted maximum of 0.5% alcohol in a drink. The effect of the legislation was to drive all forms of alcohol manufacture and distribution underground and into the hands of the mobsters.

1

'Genial' Dan Albone (1860-1906) manufactured quite a number of firsts, including the first motorcycle, the first lady's safety bicycle, the first tandem and, above all, the first agricultural tractor.

1

At Howth Castle outside Dublin, there has always been one extra place laid at table since 1576. The owner, Lord Howth, refused the pirate queen, Grace O'Malley, shelter and food, so she kidnapped his heir. The ransom demand included the terms that no stranger in search of food was ever to be turned away at a mealtime.

1

Because his father had died before he was born, King Alfonso XIII of Spain (Alfonso León Fernando Maria Jaime Isidro Pascual Antonio de Borbon y Austria-Lorena) (1886-1941) was proclaimed King of Spain at the age of just 1 month old; however, his mother ruled as Regent until he came of age at 16 in 1902. He surrendered his throne when Spain became a Republic in 1931.

1

Dress was all that Philip the Fair allowed an unmarried woman to own in 14th century France. Married women could only own 4. However, shoes were permitted, so they became the real symbols of fashion.

1

King Edward VIII succeeded to the British throne on 20 January 1936, but over the following months it was plain that his choice of wife, the twice-divorced Wallis Simpson, was unacceptable politically, socially, constitutionally and to his overseas Dominions and Colonies. Refusing to part with her, he was left with only one option, and his abdication broadcast to his subjects, on 11 December 1936, lasted just one minute.

1

Mile that the Montgolfier brothers transported some animals in a hot air balloon in 1783. A few weeks later, they sent the first humans on a nine-mile flight.

1

Month (4 March to April 4 1841) was the shortest ever US Presidency – that of William Henry Harrison. At his inauguration ceremony he insisted on standing hatless and coatless in an icy drizzle, and subsequently died of a chill.

1

The art establishment of 1878 was rocked by the libel case between London-based American artist James Abbott McNeill Whistler and Conservative critic John Ruskin, who had commented on Whistler's *Nocturne in Red and Gold* that he had "never expected to hear a coxcomb ask two hundred guineas for flinging a pot of paint in the public's face." In the ensuing court case, the defence asked the artist if he thought the two days' labour he spent on *Nocturne* worth the 200 guineas: "No", Whistler dramatically replied, "I ask it for the knowledge of a lifetime." Whistler won, but the jury awarded him only an insulting one farthing in damages, and the judge refused costs.

1

The Cyclops were one-eyed giants from Greek mythology, the most famous of whom was Polyphemus, who captured Odysseus and his crew. When asked his name, Odysseus told the Cyclops it was "Metis", which means 'Nobody'. After Polyphemus had been given unwatered wine, the drunk Cyclops was stabbed in his only eye. Hearing his cries, other Cyclops came to his aid asking who did this to him, but when he replied "Nobody" they went away!

1

The number of times a month Queen Elizabeth I (1558-1603) took a bath "whether she needed it or no". This may not sound very hygienic, but by the sanitary standards of the age was cleaner than most. On the other hand, Queen Isabella of Castile claimed to have ever had only two baths, one at birth and one on her wedding night.

2

"For two pins I would" – a phrase originating from the fact that, long ago, pins were expensive and money had to be saved up for them – hence 'pin money'.

2

Blows of the executioner's axe to behead Mary Queen of Scots on 8 February 1587 at Fotheringhay Castle, Northamptonshire. She had been imprisoned in various English castles for 19 years before her execution, ordered by her cousin Queen Elizabeth I of England.

2

Days it took a stagecoach to travel the 55 miles from London to Brighton. Today, the same journey takes about 1 hour by train.

2

In 1800, the population of Ireland was twice that of the USA. By the year 2000, the USA's population was 60 times that of Ireland.

2

Members of Parliament that the constituency of Old Sarum (with only 15 inhabitants) sent to Westminster (a so-called 'rotten borough'), while some big English cities with as many as 500,000 people had none. All this was changed by the Reform Act of 1832.

2

One of Queen Mary Tudor's main concerns of her five year reign (1553-1558) was to produce a Catholic heir, for which she needed a suitable Catholic husband. Various political and diplomatic alliances were negotiated before she succeeded to the throne, but none came to anything. However, once she was Queen, she married Prince Philip of Spain in Winchester Cathedral in 1554 within two days of meeting him. No heir appeared and her husband returned to Spain, eventually to become Philip II.

3

In May 1902, there were only three survivors from the population of 30,000 in the town of St Pierre, Martinique, after the volcanic blast from Mount Pelée. One of them was saved by being in jail, with only a tiny window facing away from the pyroclastic blast, and he was doubly lucky as he was later pardoned..

3

In the Christian Christmas tradition, three kings, wise men or 'magi', are thought to be visitors to the baby Jesus in the stable. Wrong on two counts. First, the number is a myth, as nowhere in the Bible is there any mention of the actual number. The myth probably arises from the three gifts – gold, frankincense and myrrh. Legend has given the three men names: Melchior, Balthazar and Gaspar. Second, any such visit was not necessarily at the time of birth and could have taken place up to at least two years after.

3

Legs on a 'shamble', a butchering stool. Several British cities called their slaughtering area 'The Shambles', and then some General commented that a bloody battlefield was 'like a shambles' – hence shambolic. Three towns in the UK – Nottingham, Sheffield and York – still have areas called 'Shambles'.

3

Ships whose cargo of tea was thrown into Boston Harbour. 'The Boston Tea Party' on 16 December 1773 was an act of defiance against Government from London that eventually led to America's independence.

3

The celebratory 'Three Cheers', is normally expressed by the cry "Hip, hip, hooray!" The possible origin of 'Hip' is that it stems from a medieval Latin acronym "Hierosylma Est Perditus", meaning "Jerusalem is lost". "Hooray" is probably more recent, and was more often written as "Huzza", and is just an expression of enthusiasm.

3

The number of golden balls indicating a pawnbroker's shop. Originally they were the crest of the Medici family who acted as bankers, pawnbrokers and moneylenders in the 15th and 16th centuries.

3

The shortest and last will and testament ever recorded contained just three words. The succinct '*All for mother*' was a certain Mr Dickens's last will and testament in 1806.

3

There is a myth that there are no names beginning with 'F' in the Bible. There are, in fact, three: Felix (Acts 24); Fes'tus (Acts 24, 25, & 26); and Fortunatus (1 Corinthians 16).

3

Three debtors' prisons burned down in the Great Fire of London in 1666. But the consequence of the fire was that so many people were ruined that more had to be built. Debt was a real social crisis in Britain for centuries, and many years later Charles Dickens's father was imprisoned for a £40 debt. Such ruination of lives was a constant theme in Dickens's novels.

3

Three shillings was the usual fine imposed in Anglo-Saxon times for touching the breast of an unmarried woman.

4

'The Gang of Four' were the chief members of the highly influential Chinese radical faction that helped to instigate and drive the social upheaval of the Cultural Revolution from 1966 to 1976 in Mao Zedong's Communist China. They tried to seize power after Mao's death. The members were his widow Jiang Qing and three young Shanghai politicians, Zhang Chunqiao, Wang Hongwen, and Yao Wenyuan. The coup failed, and the Gang of Four were arrested and publicly tried in 1980 to be found guilty of treason – with sentences varying from imprisonment to execution (later commuted).

4

Famous Irish beauty Elizabeth Gunning made something of a speciality of marrying Dukes, and as a result produced four sons who themselves inherited Dukedoms . Her first husband in 1752 was the Duke of Hamilton, the next the Duke of Argyll, and if it were not for a scandal involving her sister, she would have also married a third – the Duke of Bridgewater.

4th

Fourth Estate – a general term for the Press, originating from 1789 in France when Louis XVI called to Versailles the three 'Estates of Government' (first – the Clergy; second – the Nobles;

third – the Commoners). But also present was the most powerful of them all, the Press. Hence the Fourth Estate. Has time diluted the power of the Press?

4

Horsemen of The Apocalypse – the forces of man's destruction – War, Famine, Pestilence and Death – as described in the Book of Revelations in the Bible.

4.5

Feet (1.39m) – the tiny stature of appropriately–named King Pepin the Short (714-768), the father of Charlemagne.

4.7

Feet (1.4m) – the height of Etienne Lucas, the Captain of the French warship *Redoutable* at Trafalgar, and from the rigging of which one of his well-trained snipers shot Admiral Lord Nelson – himself only 5 feet 5 inches tall.

5

Lack of height appears not to have had any detrimental effect upon the reign of Queen Victoria. The Queen and Empress was only 5 feet tall and is the country's longest-reigning monarch, ruling from 1837 to 1901. She produced 6 children, 40 grandchildren and 37 great-grandchildren – amongst whom were 4 future Kings.

5

The number of stones David had for his sling shot when he faced the giant Goliath in the supposedly one-sided encounter, as recounted in the Bible's Book of Samuel. As it turned out, David only needed one shot to 'smite' the Philistine in the forehead so he 'fell on the face of the earth'. '9 cubits and a span' was the height of Philistine giant Goliath as described in the King James's version of the Bible. However, this would have made him 9 feet 9 inches tall (3m); earlier Hebrew texts had him as four cubits and a span – a much more likely 6 feet and 6 inches.

6

'Six of the best' – meaning a beating of six strokes by a cane – mostly a traditional old public school punishment.

6

Charles VIII of France, (also known as 'The Affable'), (1470-1498) was not a particularly healthy or effective King and, unusually, had six toes on his left foot. He died as the result of hitting his head on the lintel on a low doorway at the Chateau Amboise.

6

Despite the fact that the 1666 Great Fire of London ravaged a great part of the City, remarkably only six people died, the first of whom was the maid in the baker's house in Pudding Lane where the conflagration started.

6

King Henry VIII had six wives – Catherine of Aragon (divorced), Anne Boleyn (beheaded), Jane Seymour (died), Anne of Cleves (divorced), Kathryn Howard (beheaded), and Katherine Parr (survived).

6

Number of children born to the average British married couple in the 1850s.

6

The national flags that have flown over Texas: Spain, France, Mexico, the Republic of Texas, the USA, and the Confederate States of America.

6

The six Tolpuddle Martyrs were not the first trade unionists. Their actual 'crime' was to swear an oath when forming a Friendly Society, collectively refusing to work for less than 10 shillings (50p) a week. Although unions in industrial towns had been made legal in 1832, the country landowners prosecuted the six in 1834, invoking a long forgotten law of 1796 – originally passed to suppress naval mutinies. They were deported to Australia, thus becoming heroes of the nascent working class union movement.

6

There were 6 Apollo Space Missions (numbers 11, 12, 14, 15, 16 and 17) that have landed men on the moon, and as a result, 12 astronauts have walked on its surface (2 per mission).

7

The number of the theatre box in Ford's Theatre, Washington DC, in which President Lincoln was assassinated on 14 April, 1865.

7

The unusual thing about Charles – King of Sweden from 1161 to 1167 – was that he was given the ordinal 'The Seventh', yet there had been no Swedish Kings called Charles before him.

8

The number of Christian Crusades mounted by European kingdoms between 1095 and 1270 to protect Jerusalem and the Holy Land. There was a ninth Crusade in 1212 called the Children's Crusade, mounted from both France and Germany – but it was not blessed by the Pope and never arrived in the Holy Land.

9

Days – the reign of Lady Jane Grey. The dying King Edward VI wanted to prevent his very Catholic half-sister and heir, Mary Tudor, from becoming Queen and was persuaded to declare her illegitimate (as well as his other half-sister Elizabeth). His will altered the line of succession to pass to Lady Jane Grey, who was Henry VII's great-granddaughter. Edward died on 6 July 1553, and four days later Jane was proclaimed Queen. However, Mary Tudor had widespread popular support, and Jane was easily persuaded to relinquish the crown after only nine days.

9

Inches – the width of space allowed per slave on a slave ship, with only 2 feet 7 inches of headroom on a journey that could take 8 weeks from West Africa to the Caribbean. It is estimated that 10 to 12 million slaves were transported over 300 years of slavery. In return, over 10 million tons of sugar were exported from the plantations.

9

Mary Queen of Scots succeeded to the throne of Scotland on her birth in 1542 and was crowned 9 months later in Stirling cathedral. She was married first to the Dauphin of France in 1558, and, in 1559 also became Queen of France for just over a year until her husband died.

9

Months was also the age at which Henry VI was crowned King of England – he ruled from 1422 (with a Regent until 1437) to 1461 when he was deposed.

9

Muses in Greek mythology – daughters of Zeus (King of the Gods) and Mnemosyne (Goddess of Memory). Each presided over a particular art and they were companions of the Three Graces.

9.3

The Vitamin C content of an uncooked carrot is 9.3 milligrams, and that of a cooked one only 2.3mg. In the UK, people were traditionally encouraged to eat carrots to help them see in the dark. But this was a remarkable piece of British WWII wartime propaganda, to fool the Germans that British night fighter pilots were so effective thanks to their diet of carrots – rather than the real reason, the top secret radar.

9/11

11 September 2001 – the date of the Al-Qaeda attack on America, expressed in American order (month before day date) and now universally recognised as 9/11, even in countries which normally put the day before the month.

10

A Roman punishment for cowardice was when every tenth soldier was picked out from a line and then executed – hence 'decimation'.

10

Months in the original Roman Year. Hence, September, October, November and December relate to the Latin of 7, 8, 9, and 10. In 715 BC, the twelve-month calendar was introduced, based on the phases of the moon.

10

Mystery, legend and possible historical fact surround the Siege of Troy. The city was besieged by the Achaeans for ten years and eventually fell when they pretended to retreat leaving a 'wooden horse' as a gift. The Trojans were deceived, and dragged in the horse only to find that the gift disgorged 40 hidden soldiers who opened the gates, and the city was captured. Hence the origin of Virgil saying "I fear the Greeks even when they bring gifts".

10

Pounds was the price on the head of an Irish Catholic priest, dead or alive, in 1650s Cromwellian Ireland – only twice that of a wolf.

10

The really valuable commodity in the 1848 Californian Gold Rush was not the gold extracted from the ground, but the timber needed for items such as pit props, buildings and railway sleepers – all of which were worth ten times as much. The first gold find was in 1848, but by the time news circulated, the main rush of prospectors arrived the following year and were consequently known as the 'forty-niners'.

10

Tithing (from '*teogoba*' – old English for tenth) – a contribution of ten per cent of an income for religious purposes. The practice of tithing was first mentioned in the Old Testament books of Leviticus, Numbers and Deuteronomy, subsequently adopted by the Western Christian churches, and then became part of European secular law from the 8th century. After the Reformation, tithes continued to be imposed for the benefit of both the Protestant and Roman Catholic churches. However, tithes were eventually repealed in France (1789), Ireland (1871), Italy (1887), and England (1936). Tithing was never part of US law, but the congregations of some churches, such as the Mormons, are required to tithe.

11

Anne Boleyn, King Henry VIII's second wife, had this unusual number of fingers. She was beheaded for adultery, incest and treason in 1536.

11

In the London of 1900, there were 11 postal deliveries per day and 9 collections.

11

Margaret Sanger, the 20th century pioneer of contraception and women's rights, must have been affected in her career choice by her mother's 18 pregnancies and only 11 live births. Born in 1878, Margaret died in 1966 – just living long enough to see the introduction of the contraceptive pill.

12

In the 12-year rule of Alexander the Great (356BC- 323BC), he vastly expanded his kingdom from the place of his birth – Macedonia – and conquered vast tracts of Asia, having reached as far the Punjab in India. He died in Babylon just short of his 33rd birthday from mysterious causes – possibly malaria, poison, or even excessive drink.

12

King Henry VIII's foot measurement was 12 inches, and thus this length was named a 'foot'.

12

Labours of Hercules. In classical Greek mythology after Hercules mistakenly killed his wife and children, he had to serve King Eurystheus for 12 years as a penance. This required him to complete 12 seemingly impossible tasks. The successful completion of these ensured his place in the Pantheon of the Gods.

13

17th century American women gave birth to an average of 13 children.

13

Number of the main Gunpowder Plot conspirators, who on 5 November 1605, attempted to blow up the English Parliament and King James I with 36 barrels of gunpowder. Robert Catesby was the leader and Guy Fawkes the explosives expert.

14

Fourteen members of the Talbot family sat down together to dine at Malahide Castle in Ireland just before the Battle of the Boyne in 1690. All were supporters of James II, and sadly not one of them survived the battle.

17

US President Abraham Lincoln's coffin has been moved 17 times, mostly due to numerous reconstructions of the Lincoln Tomb and fears for the safety of the President's remains. The coffin has been opened five times, finally in 1901 when 23 people viewed the contents to dispel rumours that the body inside was not actually that of Lincoln. Confirmation was also needed before it was placed in a cage and covered by 4,000 lbs of cement to prevent theft.

18

Inches – the measurement of the ancient 'cubit', the length of a man's forearm.

18

Queen Anne of England ruled 1702-1714, and this last of the Stuart monarchs had 18 pregnancies. Just 5 children were born alive, and only one of these – William Duke of Gloucester – survived infancy, only to die aged 11.

20

Miles – the distance of the flight of the first practical aeroplane, the Wright 1905, on 4th October 1905 and which took 33 minutes and 17 seconds. Orville and Wilbur Wright's first flight was on 17 December 1903 and was 120 feet, less than the wingspan of a B-52 bomber.

20

The South American Mayan civilisation had a very complicated calendar. A normal month was 20 days, and a year 360 days (or 18 of their months). However, a civil year was 365 days and a religious one 200 days. They also had a formal Long Count calendar, which dates from the supposed start of their civilisation on 12 August 3113 BC, and completes its cycle on 21 December 2012 AD – when, presumably, they thought the world would end.

20 x 10

Feet (6x3 m) – the dimensions of 'Old Smith's Arms', Godmanston, Dorset – the smallest pub on one floor (although this is open to debate with 'The Nutshell' in Bury St Edmunds). Charles II stopped in Godmanston to have his horse re-shoed, and asked for a drink. The blacksmith had to refuse because he had no license, and so was immediately granted one by the King.

22

Miles of land each side of the tracks was granted by the US Government to the railways which crossed America. At one time, this added up to a huge 1/9th of the area of the United States.

23

Number of stab wounds received by Julius Caesar when assassinated by 60 Senators on 15 March, 44BC – the 'Ides of March' – a date about which he had been forewarned by a soothsayer.

23

The survivors from the 146 who were put in the 'Black Hole of Calcutta' on the night of 20 June 1756. Their cell measured just 18 feet by 14 feet.

24

24 hours and 18 minutes was how long Senator Strom Thurmond kept talking in 1957 in an unsuccessful attempt to 'filibuster' – or talk out – a Civil Rights Act in the US Senate. Thurmond served as a Senator for 48 years from 1956 to 2002, and left office at the ripe old age of 100.

24

Dollars' worth of trinkets that the Dutch paid to the Menates Indian tribe in 1626 for the island of Manahatta, one of the great bargains in history because it became New York's Manhattan.

24

The age at which William Pitt became the British Prime Minister. Son of Prime Minister William Pitt 'the Elder', he entered Parliament aged 22 and was PM from 1783 to 1801 and 1804 to 1806. This age makes 43 seem quite old for Tony Blair to have become Prime Minister in 1997.

25

Per cent – the boys between 10 and 15 who were in paid work in America in 1900.

30

George Stephenson's 'Rocket' steam engine achieved 30 mph in front of 15,000 spectators during the Rainhill Trials of 1829, to decide what form of locomotion the Liverpool to Manchester line was to have. Part of the engine's test was to pull three times its own weight and a carriage full of passengers. Stephenson built his first steam engine in 1814, but it was the 'Rocket' that set the standard for future rail travel.

30

Per cent of American states in 1900 did not allow women to keep their earnings, and 25 per cent denied them the right to own property.

33

In 1760, 33% of Britain's ships were built in America, because of the plentiful supply of cheap timber.

35

It is rumoured that President Mugabe of Zimbabwe ordered 35 custom-made, bullet-proof Mercedes at the very time that the 1985 Live Aid Concert was being performed to raise millions of pounds to help relieve the suffering of the poor of Africa. Top-of-the-range Mercedes seem to be the car of choice of dictators (Mao Tse-tung had 23) – perhaps because Rolls-Royces, Bentleys and Jaguars are more associated with colonial powers.

36

The number of existing samples of the signature of Button Gwinnett – a signatory to the American Declaration of Independence in 1776. Born in Wolverhampton, he died a nonentity in 1777, but his signature has become one of the most valuable in the world.

44

The Romans, like us, worked to a 24-hour day, split into two twelve hour segments, but these related to night and day. Thus, the length of an hour changed each day during the year. The shortest hour was 44 minutes on 21 December, and the longest 1 hour 16 minutes on the Summer Solstice. There were only two days when the hours were the modern 60 minutes – known as the equinox – 21 March and 21 September.

52

Days that the first automobile took to cross America in 1903.

52

King George I of England was 52nd in line when he ascended to the throne of Britain relatively peacefully in 1714. All the other heirs were Catholic, but this Hanoverian prince was the closest related Protestant.

53

The number of Prime Ministers, including Gordon Brown, since Sir Robert Walpole in 1721.

55

The first men who rowed across the Atlantic without the assistance of sail or canvas took fifty-five days in 1896. George Harbo and Frank Samuelsen went from America to England in an open 18-feet boat, cooked on a Primus stove and lived on tinned meat and butter, a barrel of biscuits, and eggs which they boiled in their coffee.

62

People who amazingly survived, out of 97, in the famous and horrific burning crash of the airship *Hindenburg*, at Lakehurst, New Jersey, in 1937.

63

The number of monarchs of England since the first acknowledged King, Egbert (802-839). This continuous line has only been interrupted once – between 1649 and 1659 – when the country was ruled by Oliver Cromwell, and then briefly, by his son, Richard Cromwell.

64

There were 64 Emperors of Byzantium. The dynasty was founded by Constantine the Great in 306 and ended with Constantine XI, who died when Constantinople (Istanbul) was captured by the Turks in 1453. Coincidentally, both Emperors' mothers were called

Helena. Based on the Greek-speaking element of the Roman Empire, Byzantium mostly covered the eastern end of the Mediterranean, although at times its empire stretched as far the western end and encompassed a population of 34 million. Enduring for 1,058 years, it was the longest-lasting empire in history.

80

Per cent of the world's ocean-going ships were built by Britain in 1860.

90

Feet (27.4 m) - the length of the *Mayflower* which carried the Pilgrim fathers to New Plymouth in America in 1620, transporting 102 passengers and a crew of 15-20. In comparison, today's largest supertanker's length is over 1,500 feet (457 m).

95

The 'Sun King', Louis X1V of France, came to the throne of France in 1643 aged five, and although he did not achieve full power until Cardinal Mazarin died in 1661, he was King for almost 95% of his life until he died in 1715.

101

The airship *R101* created the first air disaster when, on its inaugural international flight in 1930, it crashed into a hill at Beauvais, North of Paris, killing 48 passengers and crew. Only 7 survived.

121

Thimble Hall, Chesterfied, in England's Peak District, claims to be the world's smallest detached house – measuring 11 feet 10 inches x 10 feet 3 inches (3.3 x 1.6m) (121 square feet) (112 sq m). In 1860 it was lived in by a family of 8.

148

Years – the time a law existed which required all corpses in England to be buried in woollen shrouds in order to support the country's wool trade.

150

According to Christian Biblical legend, Noah's Ark – with all the animals – was afloat for 150 days before coming to rest on

Mount Ararat. This story is also recognised in the Islamic and Jewish traditions.

150

Miles – the distance between Stonehenge and the hills in Pembrokeshire where some of the main stones originated. There are any number of theories as to how the 4 ton stones were transported, just one of the many mysteries associated with this historic site, which was probably started around 2000 BC, and took 1,000 years to build.

151

Pseudonyms of Vladimir Ilyich Ulyanov, best remembered as Lenin (1870-1924) – perpetrator of the Russian Revolution.

179

Number of words in the 10 Commandments – as opposed to 26,253 in the European Community's rules on the sale of cabbages. The Lord's Prayer has 66 words, and the Gettysburg Address 286.

183

The weight of the Sputnik – the world's first artificial satellite in 1957 – was just 183 lbs (82.3 kg). This was put into space by Russia, was the size of a basketball, and took 98 minutes to orbit the earth.

300

The Spartan compatriots who, under King Leonidas, defended the narrow pass at Thermopylae in 480 BC, all of whom died holding up the Persian army under the great King Xerxes.

350

The pieces in which the Statue of Liberty was delivered from France to New York to be assembled like a giant kit. They arrived in 1886 in 214 wooden crates and comprised 31 tons of copper and 125 tons of steel which were the work of sculptor Frederic Auguste Bartholdi and engineer Gustave Eiffel. The statue was a present from France to celebrate the 1776 centenary of the American Revolution. However, it was erected 10 years late – delayed because the Americans took this time to raise the money required for the 27,000 ton pedestal.

441

In 441 AD, St Patrick climbed to the top of Croagh Patrick in Co. Mayo and fasted the 40 days of Lent, during which time he is supposed to have driven all the snakes from Ireland. Incidentally, as well as no snakes, Ireland has no moles or weasels.

590

The number of flights made by the airship *Graf Zeppelin* during its operating life, 1928-1937. A total of 13,100 passengers were carried without a single injury. The airship was 100 feet (30.5 m) in diameter and 110 feet (33.5 m) high (including the gondola).

967

Men, women and children who killed themselves at the fortress of Masada in 66 AD, rather than surrender to the besieging Romans.

969

According to the Bible, Methuselah, the grandfather of Noah, lived to the age of 969. A more rational explanation is that this number relates to months, which would put his age at a more reasonable 80 years and nine months.

976

The modern system of Arabic numerals did not reach Europe until it was published in the *Codex Vigilanus* in 976 AD. From the 980s onwards, this system of numbers was much promoted for general adoption by scholar and teacher Gerbert d'Aurillac, who was later the first French Pope – Sylvester II.

1,000

'A thousand years without a bath', another name for The Middle Ages – which saw very little bathing thanks to the Church's prudish attitude towards nudity, and the fact that soap was not made in England until 1641.

1,000

Feet (305m) – the depth below ground which miners could reach in 1842, having stepped on Michael Loam's 'man-engine' – the world's first lift or elevator, installed in a mine in Gwennap, Cornwall.

1,000

Guilders. In Robert Browning's famous poem, this was the fee the Pied Piper agreed with the Mayor and Corporation for ridding the town of Hamelin in Brunswick of a plague of rats. They reneged on the deal, and he then led away all the town's children. The poem is based on a legend of 1264.

1,000

Paces of a marching soldier, which became the Roman measurement of distance, the 'millia', or mile.

1,425

Theogenes, considered the greatest gladiator of ancient times, killed 1,425 opponents during his career (circa 900 BC). His preferred weapon was spiked gloves.

1,517

The number of people who died on the *Titanic* – only 306 bodies were recovered. On board was a total of 2,240 crew and passengers, of which 723 were saved, although one died on the way to New York. Other losses included 3,000 mailbags and an automobile.

1791

Belfast was once the biggest shipyard in the world. Starting in 1791, the business grew and grew, building 'the largest ship' of each generation (the most famous, perhaps, being the *Titanic*), and at its height employed 35,000 workers. By 1960 the business had dwindled, when its last passenger liner, SS *Canberra*, was launched.

1824

The year Sir William Hillary co-ordinated the first national lifeboat service, which later became the RNLI. Ten years later, there were 30 lifeboat stations, and by 1843 the 50th dedicated lifeboat was launched. By 1860, about 12,000 lives had been saved by the RNLI, and by its 100th anniversary this figure had risen to 59,975. The 100,000th life was saved in 1975.

1854

The first true manned flight was piloted in 1854, near Scarborough in Yorkshire, by the coachman of the inventor Sir George Cayley. The craft flew several hundred yards, but the terrified pilot then immediately resigned in protest, stating he had been "hired to drive a coach – not to fly".

2000

The year the supposed millennium date-related computer bug known as Y2K was meant to strike all the world's computers, because they would not recognise the next number in the ascending year sequences of 97, 98, 99. It never happened. Was this because of extreme preparation at an estimated worldwide cost of £200 billion, or perhaps by overstatement of the problem by the computer industry?

2800

The earliest record of soap-like material, consisting of ash and boiled-down fats, has been found in Babylonian clay cylinders dating from 2800 BC. However it is not certain that this was used for washing, and may have been a beauty or hair product. The Greeks bathed only for aesthetic reasons – cleaning their bodies with clay, pumice and sand and anointing their bodies with oil. According to Roman legend, soap was named after Mount Sapo, where women found that the residue of fat or tallow from sacrificed animals and ash – when combined – proved to be an effective soap. After the fall of Rome in 476 AD, bathing and the use of soap in Europe declined dramatically.

3,000

Three thousand heretics were burned in Spain by Tomas de Torquemada (1420-1498) – the Inquisitor General – during the reign of Ferdinand and Isabella. Ironically, given that his grandmother was Jewish, he also persecuted the Jews with his Spanish Inquisition.

3,618

This was the remarkable distance in miles that Captain Bligh and 18 loyal crew members sailed in a 23 foot boat for 47 days after they were cast adrift in the Pacific Ocean by Fletcher Christian, ringleader of the 'Mutiny on the Bounty' in 1789. Four of the mutineers were eventually hanged, and Fletcher Christian was killed by Tahitians on Pitcairn Island in 1793.

4,000

Oars that propelled the largest, and rather useless, ship of Egypt's King Ptolemy IV.

4,703

Deaths in London, as a result of the 'Great Smog' from 5 to 9 December 1952. (Normal rate for the time of year was about 2,000). The noxious combination of fog and excessive smoke from millions of coal fires meant for 5 days visibility was so low that sometimes people couldn't even see their feet, and at Heathrow vision was only 30 feet. Sadler's Wells had to cancel a ballet when fog in the auditorium made conditions impossible for the audience and performers. The introduction of natural gas heating and the first Clean Air Act in 1956 eliminated smog for ever.

5,000

Astrologers were retained at the court of the 13th century Mongolian Emperor, Kublai Khan.

5,500

Years ago, the first use by Egyptians of paper (papyrus).

6,000

About 85 years before Christ suffered the same fate, six thousand slaves from the Spartacus Revolt were crucified by Crassus after their defeat. Their bodies lined the Appian Way into Rome.

8,000

There are approximately 8,000 terracotta soldiers in the Chinese Emperor Ying Zheng's (260-210 BC) grave at Xi'an. The tallest is an armoured infantryman. Each face is unique, but not thought to be an exact likeness of any individual soldiers.

10,000

Pounds – the amount voted by the British Parliament to search out South American *cinchona* plants (also know as 'Jesuit's bark') and cultivate them in 1862 to produce quinine to combat malaria. Because of the amount of lives it saved, this proved to be one of the great bargains for the British Empire.

19,000

In 1939 there were 19,000 TV sets in the UK. On 2 November 1936, the BBC commenced the first TV broadcasts from Alexandra Palace – initially with regular programmes from 3pm for an hour, and then between 9.00 and 10.00 pm.

30,000

Men killed by malaria and yellow fever in the first attempt to build the Panama Canal by the 'Hero of Suez', Ferdinand de Lesseps, who was ruined by his ignorance of the humble mosquito and the efficacy of the *cinchona* tree's quinine-producing bark.

35,000

People worked in the Twin Towers of the World Trade Center, New York, together with 80,000 daily visitors. Thus it was something of a miracle that the Al Qaeda aerial attacks killed no more than 3,000.

90,000

The number of Mao Tse-tung's Communist companions on his epic 6,000 mile 'Long March' in 1934, to escape the Nationalist Chinese.

100,000

Suspected witches were killed in Germany between 1400 and 1700. In the sixteenth century, 30,000 were killed in France.

180,000

Miles of roads that connected the far-flung reaches of the Roman Empire. A message could be sent by horse relay from France to Rome in little more than a day.

1,250,000

Words in Samuel Pepys's classic 17th century diary. However, Pepys is the model of brevity compared with 20th century American diarist Rev Robert Shields (1918-2007), whose diaries run to a massive 37,500,000 words. Shields started aged 17 but soon gave up, and re-started at age 54 only to stop in 1996. Each day was often recorded in tiny 5-minute segments.

350,000

Neanderthal Man survived successfully for 350,000 years before being supplanted by our ancestor, Cro-Magnon Man. Neanderthal (named after a valley in Germany) was the first hominid to be able to exist in a cold climate (central Europe) rather than Africa. Cro-Magnon's remains were found in SW France in 1868, and he was the first hominid who can be physically related to today's humans. The only difference would be a slightly larger muscular body and skull.

2,000,000

The population of England in 1086, at the time of the Domesday Book. However, it is difficult to calculate the figure accurately because only the heads of households are recorded, and major cities as well as convents and monasteries were mostly omitted. Lincolnshire, East Anglia and Kent were the most populated regions.

7,200,000

The price in dollars for the US to buy Alaska from Russia in 1867. Although the enormous state is 25% of the US landmass, it was not initially thought to be of any great value, and the now oil-rich state was known at first, unfairly, as 'Sewald's Folly' (the US Secretary of State). Its population was 33,426; however, most were native Indians with only 430 Americans or Europeans.

8,000,000

The population of Ireland in 1845, before the ravages of the potato blight and subsequent famine, resulting in death and mass emigration. In 2007, the population of the Republic was about 4,100,000, and that of Ulster 1,700,000 – an approximate total of 5,800,000.

8,500,000

Dollars – the sum that John D. Rockefeller Junior gave to buy the United Nations its New York site.

20,000,000

People died in the influenza pandemic of 1918, 6 million more than in the devastating Great War which had just ended.

25,000,000

The number of victims of the Taiping Rebellion between 1853 and 1868 – the second highest number of fatalities in history as a result of conflict – after World War I.

60,000,000

People died of smallpox in the 17th century in Europe.

68,158,080

In 1840, the UK introduced the world's first adhesive postage stamp – the Penny Black; over 68 million were eventually printed. About 1.3 million survive, so they are not particularly valuable to collectors – now worth only between £25 and £100 each.

75,000,000

The fatalities from the Black Death, 1347 to 1351. This was three times worse than from the influenza pandemic of 1918. But if the growth of population is factored in, then in proportion the Black Death was actually 14 times more deadly.

130,000,000,000,000

Paper Hungarian pengős exchanged for one gold pengő in 1946, the worst inflation in history, even compared with Zimbabwe, in 2008. Prices rose ten times in a day.

Any Number of People

0

No President of the United States has ever been an only child. However, three might be considered 'only' as their siblings were 'halves'. Franklin D. Roosevelt had a younger half-brother, Gerald Ford had four half-brothers and two half-sisters, and Bill Clinton has one younger half-brother.

0

The thirty-third President of the United States was always referred to as 'Harry S. Truman' (1945-1953). However, the 'S' did not stand for anything at all. It was thought that his parents chose this letter to appease both grandfathers, both of whom had a name beginning with 'S'.

1/8th

American Jenny Jerome, Lady Randolph Churchill, boasted she had one eighth Iroquois Indian blood – making her son, Winston Churchill, more of a polyglot that one might have imagined.

9/16th

Of a second – the length of time CBS claimed that Janet Jackson's right nipple (along with her sun-shaped nipple ring) was on view to 90 million viewers as she sang during the American 2004 Super Bowl's half-time show. Was this an act of intentional indecency, or just an 'accidental wardrobe malfunction' as claimed by the singer? The broadcaster was fined $550,000, but the resulting legal action is still in the US courts 4 years after the event. The Federal Communication Commission's resulting brief ran to an extraordinary 75 pages.

1

Number 1, Britain Street, Southampton, was the birthplace of Isambard Kingdom Brunel (1806-1859), one of England's greatest engineers. Amongst his many contributions to Victorian life and inventions was the steel ship SS *Great Britain*, the world's first iron screw-driven liner. However, he was also famous for a number of engineering feats which are still used today, including the Clifton Suspension Bridge and the Great Western Railway between London and Bristol, which he built in the revolutionary 7 feet 'Broad Gauge', as opposed to all the other contemporary lines which were 4 feet 8½ inches.

1

Day – the length of time the great (if not greatest) English romantic author, Jane Austen (1775-1817), was engaged to Harris Bigg-Whither, who had proposed to her on 2 December 1802. She anguished overnight about her acceptance, and then declined his offer the next day. She died a spinster. Jane Austen wrote a total of 12 books, and, ironically, she wrote in a letter in 1816, 'Single women have a dreadful propensity for being poor, which is one very strong argument in favour of matrimony'.

1

In an attempt to curb the dramatic rise in population and alleviate environmental problems, China has operated an official 'one child' policy since 1979. However, there have been exceptions – in country areas parents are allowed a second child if the first is a girl, and parents are allowed to keep twins, triplets and other multiple births. There is an imbalance of more than 40 million males to females which is causing social problems. Although difficult to judge the numerical effect of this controversial policy, it has probably kept the population down by as many as 250 million people.

1

Jamie Lee Curtis is the only current Hollywood actress married to an English Lord, and thus has a title – 'Lady'. Her husband is Baron (Christopher) Haden-Guest of Saling, who achieved fame in his own right when he co-wrote the 1984 cult hit satirical film on heavy metal rock bands, *This is Spinal Tap,* also starring in it as Nigel Tufnell, a gormless rock star.

1

Not only was Isaac Newton (1642-1727) a brilliant physicist and scholar, he was also the Member of Parliament for the constituency of Cambridge University in 1689, and then from 1701 to 1702. However, he did not contribute much to British parliamentary life, and only spoke once in the House of Commons

- merely to ask for the windows to be opened.

1

Penny – this paltry amount was offered to the eccentric, shabbily dressed, but vastly rich Duke of Northumberland by a lady who asked him to carry her bag, saying "No doubt it is the first honest penny you've ever earned". "Indeed it is", he replied.

1

Pilot's Licence number 1 in the UK was granted to John Brabazon, who made the first flight in Britain in May 1909 – 500 yards in a Wright Flyer made under licence by Shorts. Some years later, he was the first man to fly under London's Tower Bridge.

1

The only Head of State of a major country in 1789 who was not a monarch – George Washington, President of the USA.

1

The steamy sex video, *One Night in Paris*, is the 2004 privately-made video of Paris Hilton's night with Rick Salomon. The video found its way on to the Internet and thus ensured the notoriety of its eponymous star – even though she sued her partner for releasing the footage without her permission. The video is very explicit, but in one scene Paris at least ceases 'what she is doing' to answer her mobile phone.

1.5

It was scurrilously rumoured that the length of Napoleon Bonaparte's penis was only 1.5 inches (3.8 cm). He was 5 feet 6 inches (1.7 m) tall, and, true or not, it is said that the autopsy revealed he had undersized genitals.

2

The number of cigar types named after personalities – a (Sir Winston) 'Churchill' is normally 7 inches (17.8 cm) long with a 3/4 inch (1.9 cm) diameter, and the (Earl of) 'Lonsdale' is likely to be 6 to 6 3/4 inches (15-17 cm) long and slightly thinner than a 'Churchill'.

2

Words – a young woman bet her friends that she could get more than two words out of the famously taciturn US President, Calvin Coolidge (1872-1933). "You lose", he replied.

3.3

About thirty thousand Americans commit suicide each year (almost one every 15 minutes). However, sadly, the incidence of suicide is over three times more prevalent amongst young Native Americans compared with the rest of the population.

4

Per cent – the disastrous mark given to co-author Donough O'Brien in his school Advanced Maths exam. He had, in fact, been let off lessons in this subject because he was so hopeless. But as the school had paid the entry fee, it made him sit the exam, and so for two hours he looked blankly at the questions, which by then he did not even understand, with the inevitable result!

4

The amount of attempts at O Level Elementary Maths made by co-author Anthony Weldon – making it curious to be doing a book on numbers!

4

When Queen Elizabeth II was crowned in 1953, the Archbishop of Canterbury, Dr Geoffrey Fisher, handed her the four 'symbols of authority' – the orb, the sceptre, the rod of mercy and the royal ring of sapphire and rubies. He then placed St Edward's Crown on her head to complete the ceremony.

6

There are six Presidents of the United States who served in the US Navy – John Kennedy, Lyndon Johnson, Richard Nixon, Gerald Ford, Jimmy Carter and George Bush senior.

10

The reigns of English monarchs witnessed by 'Old' Thomas Parr, who supposedly lived to 152. Said to have been born in 1483, he had an affair aged 100 and fathered a child out of wedlock. After the death of his wife, he remarried at 122. Sadly, his visit in 1635 to Charles I 'cut his life short' due to London's polluted air, and he was buried in Westminster Abbey by the direction of the King.

12.5

Ex-MP, author and raconteur, Gyles Brandreth, broke the record for the longest after-dinner speech when he spoke for a marathon twelve and a half hours in order to raise funds for the National Playing Fields Association.

11

Franklin Delano Roosevelt was related to eleven other Presidents – 5 by blood, and 6 by marriage.

11

Paris Hilton's unusually large shoe size.

11

The number of jobs that the 'completely disabled' John Gaster was holding down in 2007, including kitchen porter, carer, support worker and shop assistant. Claiming £17,000 in Disability Allowances, he was incredibly only found guilty of 'not revealing his changed circumstances'.

13

When Anne Frank's parents gave her a diary bound in red-check for her thirteenth birthday in 1942, little did they realise how important this book would become. Soon afterwards, the Frank family went into hiding to avoid the Nazi persecution of the Jews in German-occupied Amsterdam. For two years, she confided her closest thoughts to her diary and recorded a remarkable account of her family's amazing survival, living silently in a hidden attic until they were betrayed. Although she died in a concentration camp her diary survived, and was first published in 1947. It has gone on to sell millions of copies worldwide.

14

One of the greatest symbols of a husband's love for his wife was the building of the Taj Mahal mausoleum in Agra, India, by the 5th Mughal Emperor, Shah Jehan (1592-1666), for his third and favourite wife, Mumtaz Mahal ('Jewel of the Palace'). She died in 1631 aged 38, while giving birth to their fourteenth child.

15
In 1965, the opinion of LSD guru Timothy Leary was that fifteen years was all that the United States had left before collapsing. When challenged about this in 1980, he replied "What is time?".

15
In 1968, artist and leading figure in the Pop Art movement, Andy Warhol (1928-1987), predicted that "In the future everybody will be world-famous for fifteen minutes".

16
The age of consent for heterosexual and homosexual men and women is 16 in England, Wales and Scotland, and 17 in Northern Ireland and the Republic of Ireland. The lowest in Europe is 13 in Spain (provided there is 'no deceit' to gain consent), and Turkey and Malta have the highest at 18. In the USA, it varies from state to state and between 16 and 18.

17
Inches – probably the smallest genuine fashionable 'wasp-waist' (no doubt heavily corseted) of the French dancer Polaire (1874-1939). Concepts of fashion have fortunately changed over the years – the 1950s epitome of womanhood, Marilyn Monroe, was 37-23-36; the 60s supermodel Twiggy was 32-22-32. Later models such as Naomi Campbell are 34-24-35, Claudia Schiffer 36-24-35, and Kate Moss is a slightly more slender 32-23-32.

20
The Nashville, Tennessee, mansion belonging to Al Gore (former US Vice-President), global-warming and energy conservation guru, has 20 rooms with 8 bathrooms. At one time, it apparently, and ironically, used 12 times the energy of the average American home.

24
There are twenty-four hereditary non-Royal English Dukes. In addition, some hold more than just one Dukedom – Abercorn (created 1866), Argyll (1701), Atholl (1703), Beaufort (1682), Bedford (1694), Buccleugh (1663) (& Queensberry (1664), Devonshire (1694), Fife (1900), Grafton (1675), Hamilton (1643) (& Brandon (1711)), Leinster (1766), Manchester (1719), Marlborough (1702), Montrose (1707), Norfolk (1483),

Northumberland (1766), Richmond (1675) (& Gordon (1876) and Lennox (1675)), Roxburghe (1707), Rutland (1703), St Albans (1684), Somerset (1547), Sutherland (1833), Westminster (1874), and Wellington (1814).

35

At the age of 35, Civil Rights leader Martin Luther King Jr (1929-1968) was the youngest-ever recipient of the Nobel Peace Prize. When notified of his selection, he announced that he would turn over the prize money of $54,123 to the furtherance of the Civil Rights movement.

36

Dr Samuel Johnson (1708-1784) – renowned scholar, poet, Parliamentary reporter, essayist, moralist, critic and lexicographer – once drank 36 glasses of port without stirring from his seat, a remarkable feat even by the typically bibulous standards of the 18th century.

36

Number of William Shakespeare's (1564-1616) plays, plus two additional ones *(Henry VIII* and *Two Noble Kinsmen*) written with his collaborator John Fletcher.

37

It took Frenchman Louis Bleriot, the first man to fly the English Channel, just 37 minutes to cross on 25 July 1909. A year later, Charles Rolls, of Rolls-Royce fame, was the first to fly both ways.

38

Before the future US President Jimmy Carter was born in a hospital in 1934, the previous thirty-eight US Presidents were all born at home.

40

American motorcycle stuntman Evel Kneivel (1938-2007) broke a total of 40 bones as a result of his daredevil acts, and subsequently spent a total of 3 years of his life in hospital.

40

Even forty years after Prince Albert died, each evening Queen Victoria ordered that his evening clothes be laid out at Windsor.

40

Forty per cent of American Presidents have some Irish ancestry.

40

Inches (3 feet 4 inches) (1 m) – the height of Tom Thumb (1838-1883) the celebrity circus character, at the time of his death. He was born a normal size, but by the age of 5, when he made his first performance for circus owner T.P. Barnum, he was 25 inches tall and had not grown at all from the age of six months old. He finally started to grow very slowly, and at the age 40 he was still only 2 feet 11 inches. However, despite his tiny stature, he made a huge fortune and name for himself.

46

It is claimed that the remarkable romantic novelist Dame Barbara Cartland, (1901-2000), received at least forty-six proposals of marriage. Belonging to an old-fashioned and somewhat innocent generation, she is supposed to have broken off her first engagement, to a Guards officer, when she learned about sexual intercourse. However, she was married twice and was step-grandmother to Princess Diana. Known mostly for her prodigious writing – 723 books with a billion copies sold worldwide – she also took an early interest in gliding and was a pioneer woman racing driver in the 1930s at the rather exclusive racing circuit at Brooklands (advertising slogan: *The right crowd and no crowding)*, where there is still a room dedicated to her.

59

Famous French artist Henri de Toulouse-Lautrec (1864-1901) was only 59 inches tall (4 feet 11 inches) (1.5 m). Deprived of the physical life that a normal body would have permitted, Toulouse-Lautrec lived completely for his art. He moved to Montmartre in Paris, which was the centre of the French bohemian life of dance halls and nightclubs, racetracks, and prostitutes which he brilliantly depicted on canvas or made into lithographs. His dissolute life led to a premature death from alcoholism and syphilis, with his last words reportedly being *"Le vieux con!"* ("Old fool!").

63

Years that Queen Victoria reigned, from 1837 to 1901. Louis XIV, 'the Sun King', achieved eight more by ruling from the age of 4 to 76, when he died in 1715.

72

Alexander Graham Bell, the inventor of the telephone, was seventy-two years old when he set the then world motorboat water speed record in 1919, of 70 mph.

74

John Faulkner, the world's oldest jockey, rode his last race when he was seventy-four years old. He died in 1933 aged 104, having fathered 32 children – a truly sporting fellow!

77

John Glenn became the world's oldest astronaut at seventy-seven when he went into space in 1998. He first became an astronaut 36 years earlier, in 1962, when he manned the USA's first orbital mission, and, all told, has logged 218 hours in space. In between, he was US Senator for Ohio from 1974 to 1999.

90

According to a Royal Commission in 1868, a remarkable 90 per cent of women in Scotland were pregnant on their wedding day.

99

The owner of the world's largest chain of jewellery shops, Gerald Ratner, told the Institute of Directors at their 1991 conference that he could sell a pair of earrings for 99 pence, but added the unfortunate phrase "Less than a Marks & Spencer prawn sandwich, but wouldn't last as long". This ill-considered joke, together with his remarks about his sherry decanters being "crap", lost him his business empire.

100

The first reward offered for Dick Turpin, later the infamous highwayman, was £100 for the rather unromantic crimes of robbing farmhouses and torturing the occupants. He was eventually executed in York on 7 April 1739.

130

During her reign (1952-) Queen Elizabeth II of England has visited 130 countries (Australia 16 times, Canada 24, and New Zealand 10), launched 23 ships, sat for over 140 portraits, been Patron of 620 charities and organisations, and on a personal level been 'mistress' to 30 corgis.

200

Pounds that Sir Henry Morgan, the buccaneer, was awarded for defamation when a newspaper called him a 'pirate'.

400

In the second half of the 19th century, the four hundred who were accepted as being New York's 'Fashionable Society' were so defined because this was the maximum number that could fit into the ballroom of the city's tyrannical social *grande dame*, Mrs Caroline Astor.

1,060

The actual number of pairs of shoes that Imelda Marcos, known as the 'Steel Butterfly', and the notoriously spendthrift wife of Ferdinand Marcos (President of the Philippines from 1965 to 1986) insisted she owned – as opposed to the rumoured '3,000' pairs. Many of them went on display in a Manila museum in 2001. In a US court case brought on charges of corruption, she sighed, "I get so tired listening to a million dollars here, a million dollars there. It's so petty!".

1952

The longest-running play, *The Mousetrap,* was first put on in London in 1952. The author, Agatha Christie, wrote a total of 15 stage plays and 68 crime books and stories. She also penned 6 romantic novels under the name Mary Westmacott.

4,862

There was a deliberate 19th century policy in America to eliminate the bison and thus starve the Native American tribes, in order to make way for the westward expansion of the railroads. In one year 'Buffalo Bill' Cody alone killed 4,862, and he, with others, helped to reduce the bison population from 13 million to almost none.

10,000

The number of Royal Princes of the House of Saud – the rulers of Saudi Arabia.

15,000

Dresses found when Elizabeth I of Russia died in 1762. She sometimes changed 3 times in an evening.

27,000

The probable feet above sea level on Mount Everest reached by climber George Mallory in 1924 before he died. There is considerable debate as to whether he actually reached the top – if so, he was the first to climb the world's highest mountain.

42,000

Between 1908 and 1965 it is reputed that Winston Churchill had forty-two thousand bottles of his favourite champagne, Pol Roget, opened for him.

400,000,000

Dollars, the fortune that Scottish-born steel magnate, Andrew Carnegie, gave away before his death, including funding hundreds of libraries and Carnegie Hall.

430,000,000

Charles Osborne of Iowa, USA, started hiccupping in 1922 and only stopped 68 years later in 1990. During that time it is estimated that he hiccupped 430 million times. (On average, humans probably hiccup only around 2,330 times in a lifetime). His affliction didn't stop Osborne from leading a normal life, becoming something of a TV personality or prevent him from getting married 5 times. He died one year after he stopped hiccupping.

531,326,842

Dollars that John D. Rockefeller gave away to philanthropic causes, after a ruthless career creating Standard Oil, during which he did not display much soft-heartedness.

MONEY, MONEY, MONEY.

0

Purchasers of Henry Ford's Model T were offered 'any colour of car, provided it was black'. This zero option only lasted between 1913 and 1927. Prior to 1913, grey, red or green were the other options. The Model T was the foundation of the Ford car empire, and over 15,000,000 of them were produced, sales only to be exceeded by the Volkswagen 'Beetle' in 1972.

0

The amount of Stilton cheese made in Stilton village, Cambridgeshire. The cheese was actually produced in Leicestershire, but was made famous by the keeper of the 'Bell Inn' in Stilton, the much-used coaching stop on the Great North Road.

.04

In 1947, the price of a pint of Guinness in Ireland was under .04 cent (.05p), by 1969 it was .20 cent, by 1989 €1.1.87, by 1999 €2.74, and by 2007 it had increased considerably to €4.03 (£3.18).

1/10th

The British insurance industry employs 330,000 people and accounts for 1/10th of the UK's GDP. The industry manages £1,085 million on behalf of clients such as life and pension businesses, which in turn accounts for 17% of the London Stock Market's business.

1/3rd

The basic attraction of nuclear fuel is its low fuel costs – typically about 1/3rd of a coal-fired generator, and 1/5th of a gas combined-cycle plant.

1

One peal of the *Lutine* Bell in the Lloyd's of London insurance market used to announce a shipwreck of an insured ship, or other bad news. (Two peals meant good news or the arrival of an overdue ship). Originally the ship's bell on a French frigate which surrendered in 1793. Six years later, as HMS *Lutine*, the ship sunk with a valuable cargo of silver and gold bullion which was insured at Lloyds. After many attempts to retrieve the cargo, the bell was saved and placed in the Lloyds underwriting room. Despite heavy losses at sea during WWII, it was rung only once – on the sinking of the *Bismarck*. Although now mostly only used for ceremonial occasions, the Lutine Bell has been rung recently on receipt of bad news, such as the 9/11 terrorist attack, the London bombings and the Asian tsunami.

1

'Buy one, get one free', a marketing ploy better known by the acronym 'BOGOF'.

1

'Meals for One', a great marketing failure, because it reminded people of loneliness. Changed to 'TV dinners'!

1

"A1" was what King George IV proclaimed in 1824, when he tasted his chef Henderson William Brand's spicy sauce. In 1831, Brand went into commercial production with his sauce, with 'A1' as the product name. After years of various trademark disputes and different owners, it was introduced into the United States in 1895, and subsequently became the best-selling product it is today.

1

Finland's position in the league table of mobile phone ownership – mostly thanks to its own company Nokia, (which has 39% of the world market). There is even an annual mobile phone-throwing competition at Savonlinna. The longest throw so far is 271 feet (82.55 m).

1

In 1890, one per cent of the American population owned more wealth than the other 99 per cent combined. Even today, 8 per cent own more wealth than the other 92 per cent.

1

In 1903, Thomas Alva Edison announced his new, more efficient and rechargeable battery to great fanfare. Despite extensive testing, he had to refund $1 million to buyers of his batteries after one was found to be defective. Edison was an American inventor who, singly or jointly, held 1,093 patents.

1

One ounce of gold can be beaten out into a thin film measuring 100 feet square and 1/282,000th of an inch thick.

1

One shilling was charged by Thomas Cook (founder of the eponymous travel company) for his first package tour, which was from Leicester to Loughborough for a Temperance meeting in 1841.

1

Such was the perceived value of tulips in Holland in 1635, that just one bulb was exchanged for 1,000 lbs of cheese, 4 oxen, 8 pigs, 12 sheep, a bed and a suit of clothes. Even big houses were exchanged for single bulbs during the 'Tulipomania' bubble, which, when it collapsed in 1637, nearly wrecked the Dutch economy.

1

Year – Thomas Edison is normally credited with the invention of the light bulb in 1876, but Sir Joseph Swann had preceded him by at least a year. At first they were rivals, but eventually became partners in the 1880s.

1.05

This was the value in decimal currency of a 'guinea' – £1 and 1 shilling, consisting of 252 old pennies. Introduced on 6 February 1663, the guinea was the first British machine-struck gold coin, which varied in value depending upon the price of gold. Cancelled as a coin on the introduction of the £1 sovereign in 1816, it continues as a named unit of currency (even after decimalisation in 1971) for the sale of livestock, and particularly racehorses.

1.5

Tons – this was the possible modest weight of a future computer predicted by *Popular Mechanics* in 1949, compared with the reality of the tiny size and weight (yet vast computing power) of the modern computer, and its driving force, the microchip.

2

2CV – the famous French 'Deux Chevaux' (2 horse power) Citroen car was designed for cheap, simple, cross-country use by French farmers. Between 1948 and 1990, nearly four million were produced.

2

Tons of South African ore are mined for every one ounce of gold.

2.4

The percentage of millionaire households is the same in both the UK and Ireland – 2.4%. The highest is in the United Arab Emirates with 6.1%. The USA has the highest number of dollar millionaire households with 4,585,000; next is Japan with 830,000, and then the UK with 610,000.

3

Miles – the internationally accepted standard minimum distance between commercial aircraft coming in to land, although this may vary. If a smaller plane is coming in behind a big commercial airliner, then the air turbulence requires a greater distance.

3

The third generation of fibre optics will process 10 trillion bits of information, the equivalent of transmitting 150 million phone calls and 1,900 CDs each second.

4

Only four women have appeared on British banknotes – Queen Elizabeth II, Victorian nurse Florence Nightingale, prison reformer Elizabeth Fry and the symbol of the country's national identity, Britannia.

4

The four 'Cs' that give the value of a diamond: carat (universal unit of weight), clarity, colour and cut. The Golden Jubilee is currently the world's largest diamond at 545.67 carats. However, one of the world's most famous, the Koh-i-noor ('mountain of light'), was originally owned by Mughal emperors; when it was given to Queen Victoria in 1850 it was 186 carats, but to show it better, it was re-cut to 106 carats and then fitted into the coronation crown of the British Queen Consort.

4-8-8-4

The wheel configuration of the largest steam locomotive ever built, Union Pacific's massive 600-ton 'Big Boy'. Only 25 were built, and they ran from 1941 to 1959 producing 6,000 horsepower at 80 mph, pulling 15,000 ton freight trains.

5

One of the most popular and commercially successful perfumes in the world, 'Chanel No 5', so named because Coco Chanel had rejected the first four fragrances presented to her. This was the perfume which Hollywood film star, Marilyn Monroe, seductively claimed "was the only thing she wore in bed".

5

Per cent – the commission which the famed extrovert and entrepreneurial businessman Nubar Gulbenkian (1896-1972) levied on his various oil deals, accruing him enormous wealth and giving him the name 'Mister Five Per Cent'. He claimed his specially-built replica of a London taxi could turn on a sixpenny coin, adding patronisingly, "whatever a sixpence may be".

5

Pounds of beef that caterers, the Menches brothers, were forced to try out, because no pork was available at the Erie County Fair in 1885, at Hamburg, New York. They called their popular new sandwiches 'hamburgers'.

5

The number of centres where Coca Cola's secret formula is held and from which the base syrup is then distributed to bottlers all over the world. The ingredients included extract of cocaine until this was banned in 1905. Originally formulated by Dr John Pemberton in 1886, in its first year Coca Cola sold 3,200 servings – a somewhat modest amount in comparison with the Coca Cola company's modern turnover. Today the company has a market capitalisation of $123,526.5 million, and employs over 90,000 people.

5

The total future world market for computers was predicted as only 5 in 1943 by the highly respected businessman, Thomas J.

Watson of IBM. Despite the fact that this prediction did not come true, it did not affect IBM's commercial success or the subsequent explosion in numbers of personal computers. Some modern researchers believe Watson's prediction of five computers may actually be four too many, and that the world could manage with just one supercomputer.

7

Introduced in 1982, the British 20 pence coin has 7 sides – each of which is equilaterally curved so that the coins roll freely in slot machines.

7

Per cent, or about 90 billion euros of Italy's Gross Domestic Product is generated by the Mafia and organised crime.

7

There were originally seven secret ingredients after which Charles Grigg re-named his carbonated drink, 7Up, wisely changing it from 'Bib-Label Lithiated Lemon-Lime Soda'. The product was launched in 1929, just two weeks before the Wall Street Crash.

7/11

Name of a popular local shop chain in America, denoting its stores' opening and closing times.

8

American W.G. Peacock used to sell individual vegetable juices, but with only moderate success. However, in 1933 he combined all 8 vegetables into one, packaged the new drink as V8, and, such was its success that in 1948 it was bought by the present owners, Campbell's Soup.

8

In 1986 Kuwait Petroleum International very successfully re-branded itself and made a new identity under the onomatopoeic brand name Q8.

8

Miles per hour – the 'reckless speed' of Walter Arnold's car in 1896. A policeman pedalling furiously on a bicycle finally caught up with him, and Arnold was subsequently fined 1 shilling for exceeding the 2mph speed limit in the built up area of Paddock Wood. He went on to build the first British petrol-driven car, and fitted the world's first self-starter.

8

Piece of Eight – also known as the Spanish dollar. A silver coin worth 8 reales, first minted in the Spanish Empire in 1497. Only ceased to be legal US tender in 1857. Associated with pirates and books such as *Treasure Island*, with Long John Silver's parrot reciting "Pieces of Eight" (parodied in Terry Pratchett's *Going Postal* with Reader Gilt's parrot crying "12.5%!").

8

Pints of milk needed to make one pound of Cheddar cheese. Britons eat 650,000 tons of cheese a year, of which 450,000 tons are Cheddar.

8

The number of applicants who were rejected as potentially unsuitable members of the first formalised London Stock Exchange in 1801. In 1802, on renewal of membership, 498 dealers were confirmed members but 17 rejected for not being acceptable. Irish Stock Exchanges, in both Dublin and Cork, have roots dating back to 1793. In 1973, the Irish Exchange became part of Great Britain's International Exchange, but then became independent in 1995.

10

Dollars a page, agreed by the two rather broke artists who created *Superman*, Joe Shuster and Jerry Siegel, without keeping any other rights. Harry Donnenfeld of Detective Comics Inc was making $100,000 a year out of them by 1940, and the film and TV takings don't bear thinking about.

10

Every ten minutes the British banking system processes 25,000 cheques, 102,000 automatic credits and 123,000 plastic card payments. In the same time, bank ATMs will have paid out £3 million in cash.

12

Seconds – the duration of the Wright brothers' first powered flight in 1903. It also took 12 seconds to telegraph the news of this remarkable event. Kittyhawk (North Carolina) is often the place associated with this major aeronautical achievement – however, it was in fact only the nearest post office, and thus it was from there that the news was sent out. The flight actually took place at Kill Devil Hill, four miles away.

12

Traditionally, a dozen roses send a message of love. But why a dozen? Although the origins are unknown, the flower shops are eternally grateful for the business.

13

'Baker's Dozen' – of uncertain origin, but the extra loaf over 12 is thought to be the baker's profit.

15

Miles per hour – the average speed in 1830 between London and Bath of Sir Goldworthy Gurney's first self-propelled motorised steam carriage – the forerunner of the national bus service.

15

One of the world's oldest financial markets, the London Stock Exchange, started life in the City's coffee houses, and in 1689, there were initially 15 major traded joint stock companies. With the increase in international trade this had grown to 150 by 1695. The current London Stock Exchange lists 3,305 companies. The coffee house has also changed with its first formal trading room in 1773 and first dedicated building in 1802. Women were not allowed to trade on the Floor until 1973, and 1985 heralded the major changes of electronic trading, merging of brokers and jobbers and the ownership by outside corporations.

15

The 'Light Fifteen' was the elegant and distinctively designed revolutionary front-wheel drive Citroen first produced in 1935, and beloved by fans of the fictional French detective Maigret whose car it was.

15

The first ever air travel insurance was bought on 23 October 1922, when Mr Clanahan of Cornwall was flying between Manchester and London. At 1,500 feet, he purchased £500 of cover for 15 shillings (75p) from an insurance broker who was on the same flight.

20

Minutes, the time it would take a supertanker to stop at 16 knots.

20

The number of standard 75cl bottles in the biggest size of champagne bottle – a Nebuchadnezzar. The other sizes are Magnum (2 bottles), Jeroboam (4), Rehoboam (6), Methuselah (8), Salmanazar (12) and Balthazar (16).

25

The patent for Watts' innovative steam engines lasted 25 years, from 1755 to 1780, thus giving him an amazing monopoly on this very profitable new invention. His customers were charged on the basis of how many horses his steam engines saved them – thus the origin of 'horsepower'. Watt's schoolmaster was somewhat wrong when he called the young James Watts "dull and inept"!

25

The wage offered to riders on the Pony Express mail service was $25 per week in one of the frankest advertisements in history; 'WANTED. Young, skinny, wiry fellows. Not over 18. Must be willing to risk death daily. Orphans preferred'. However, of the 186 riders, who included the 15 year old Buffalo Bill Cody, only one rider and one mail bag was ever lost. The service used 400 horses, changing them every 10 miles at 165 stations over a distance that, both ways between St Joseph, Missouri and Sacramento, California, was 2,000 miles. Each rider carried 20 lbs of letters on especially light paper, but their employment came to an end in 1861 with the introduction of the telegraph and the railroads.

30

It takes thirty mulberry trees to produce one kilogram of silk from the silk larvae that feed on their leaves. Over a three to eight-day period, the larvae perform about 300,000 figure-of-eight movements to form their silky cocoons, and, when these are harvested, about 200kg of cocoons will produce 40kg of silk.

30

In 1900, thirty per cent of the cars in New York, Chicago and Boston were powered by electricity.

35

Imperial gallons (42 US gallons) that make up a barrel of oil. Each barrel is refined into 19 gallons of petrol with the rest becoming jet fuel, heating oil and other petroleum-based products. UK beer barrels contain 36 gallons.

40

WD 40. The universal oil that was invented in 1953 in San Diego, California, by a 3-man company that made rust preventative products for the aerospace industry. The 40th version of their **W**ater **D**isplacement product was the only one that really worked – hence WD40.

41

At the current rates of consumption, the world has only 41 years of oil reserves, 61 of gas and 267 of coal.

48

Miles – the length of the first 'transcontinental' railway, completed across the Panamanian Isthmus in 1855. This was 14 years before the more famous 'Golden Spike' ceremony in Utah in 1869, which celebrated the joining of the Central Pacific and Union Pacific railroads for the first US transcontinental railway.

51

Brand name of the most famous French pastis drink.

51

The Parker Pen Company's most famous product, 'The Parker 51', was created in 1948 to celebrate its 51st anniversary.

53

Fifty-three per cent of 1,583 Chief Executives polled in the USA are firstborn children.

54

In 1963, fifty-four per cent of the UK population invested directly in Stock Exchange quoted equities. By mid 1990s, this figure was 17.7%. However, with the growth of investment trusts and other such financial vehicles, the actual number of investors grew from 3 million in 1979 to 12 million in 1997.

57

Heinz Varieties. Henry J. Heinz successfully started with bottled horseradish (1869) in clear glass to show there were no cheap fillers, and went via sour gherkins to tomato ketchup in 1876. (Tomato ketchup was not distributed in the UK until 1896). Heinz was inspired by seeing another advertisement for 21 varieties of shoes, so he decided that he too would have a number of varieties. When '57 Varieties' was coined, there were already more than that, but he simply liked 'the ring of the number'.

60

Modern high technology airships are considerably more fuel efficient than conventional alternatives – 60% less fuel is consumed than by the equivalent fixed wing aircraft, and 85% less than helicopters.

64

"Sixty-four kilobytes ought to be enough for anyone" – was the opinion of young Bill Gates in 1981 about the future requirement of computer capacity.

69.5

This was the depth in feet of the first oil well, drilled by Edwin Drake at Titusville, Pennsylvania in August, 1859. From then on, Pennsylvania produced half the world's oil, until Texas boomed in 1901.

70

As a result of champagne's secondary fermentation, the pressure under which it is stored in a bottle is between 70 to 90 lbs per square inch, thus requiring thicker glass than ordinary wine bottles. This pressure is two to three times greater than that of a car tyre.

70

Scotch whisky and Irish whiskey are normally sold at 70 degree proof. However, when they first come out of the still, they are 115-120 degrees proof and this is then diluted down on bottling. (Proof is twice the per cent of alcohol by volume).

73

The commercially successful American multiple-shot lever-action rifle, the 'Winchester Seventy Three', was produced in 1873 – the same year as the Colt.45 Peacemaker pistol. Both weapons have entered American cowboy folklore by being called "the guns that won the West". Winchester sharpshooters could squeeze off 15 accurate shots in 10 seconds.

75

In a company/work environment, the amount of corporate email traffic compared with non-office and personal traffic is estimated as 75%. Overall, 65% of all types of email traffic are spam, whilst this figure is only 10% for SMS on mobile phones.

89

North America has an 89% rating for the efficiency and speed of its internet connections, Australia is the next highest with 80%, then Asia and Europe with 75% – and South America has the lowest with 63%.

103

The well-known Spanish brandy, 103, is distilled by the Spanish Osborne family, whose ancestor Thomas was from Exeter and founded the company in 1772.

110

Since the collapse of Communism, an enormous number of fortunes have been made in Russia. It is reckoned that there are 110 dollar billionaires and at least 100,000 multi-millionaires.

150
Years since the longest-running fashion dynasty – The House of Worth – was founded in 1858 by Charles Worth. Originally born in Lincolnshire in 1825, Worth re-located to Paris to work in the fashion industry. The success of his company was continued by his sons.

250
Amsterdam has 250 officially registered brothels. Most of the prostitutes are in the Red Light district, waiting for custom in the 300 or so 'shop windows'.

300
In 1914 there were over three hundred national American automobile manufacturing companies; now there are only three main ones left – Chrysler, Ford and General Motors.

501
The origin of Levi Strauss 501 jeans is not absolutely certain, as the company's records were lost in the 1906 San Francisco earthquake and fire. However, the name certainly related to the type of rivets used to give extra strength to the product – a patent for which was issued to Strauss (1829-1903) in 1873.

555
The cigarette brand, State Express 555, was launched in 1895 and named after the famous train – New York Central's 'Empire State Express'. The cigarette brand has subsequently been promoted extensively through motor-racing sponsorships.

707
The Boeing 707 airliner became the epitome of the jet age when it was introduced in 1958, but only after the success of its competitor, the world's first jet airliner – the British de Havilland Comet – was restricted by crashes caused by metal fatigue. Over 20 years, Boeing made 1,010 707s.

747
...and some more about Boeing. In 1970 the first jumbo jet, its 747, revolutionised air travel for both passengers and freight. It was two and half times the size of a Boeing 707, and one of its

engines had more power than all four of a 707's; the 747 cost $24 million and could carry 600 passengers. After a number of new models and variations, the latest model, the 747-8, costs $285-300 million. Up to December 2007, 1,396 747s have been sold and over 125 more are on order.

911

The Porsche 'nine eleven', rear-engine rear-wheel drive, was launched in 1963, and is still going strong, and for some, represents the epitome of a classic sports car.

914

The revolutionary Xerox copier model 914 launched photocopying in 1959, dooming the typewriter's carbon copy. (The Xerox Company was originally called the Haloid Company in New York, but changed its name to one based on the word 'xerography' – meaning 'dry writing').

1001

The traditional, but still very effective carpet cleaner 1001 came to fame with the advertising jingle "1001 cleans a big, big carpet for only half-a-crown".

1,500

Although it thought that sausage-making originated in 3000 BC, sausages are mentioned in both China and Greece in 500 BC. The word derives from the Latin word 'salsus', meaning salted or preserved. Perhaps it is the Germans who have the most variations – 1,500 approximately. The UK has about 400 varieties of sausage.

1,400 – 1,600

Degrees Fahrenheit, (760-871 °C) – the temperature at which diamonds break down, contrary to the popular belief that 'diamonds are forever'.

3,106

Carats in the Cullinan Diamond, which was cut into 105 stones, two of which are in the British Crown Jewels.

4,000

Passengers for which Brunel's liner was designed; twice as many as those in the *Queen Mary* 75 years later. His *Great Eastern* could carry enough coal to circumnavigate the globe.

4711

The world famous eau-de-Cologne originally made by a French émigré friar in 1794, and named after the street number of the perfumery shop. Cologne's water was thought to have powers to ward off the plague, and was developed by Giovanni Maria Farina (1685-1766). Cologne has only 2-5% of concentrated essential oils, while modern perfume has 20-40%.

8,601

Number of diamonds encrusted in Damien Hirst's sculpture of a skull, titled 'For The Love of God' and sold for £50,000 million. In the centre of the forehead is a 52-carat pear-shaped diamond.

30,000

Behind Japan, Ireland is ranked the second wealthiest nation per capita in the top OECD nations, with 30,000 millionaires. Much of this wealth has been created in the last ten years, growing by 350% thanks to a booming economy and rising property prices.

32,000

'Speakeasies' or illicit bars, that thrived during 'Prohibition' in New York alone. These replaced 15,000 legal saloons. There were no less than 100,000 speakeasies in America. No wonder the anti-drink experiment failed.

50,000

The number of millionaires in 1991 in the UK. With the increase in value of shares and property, by 1995 this figure was 81,000, and by 2004 had risen to 230,000. It is forecast that by 2020 there will be 1.7 million millionaires. The average will be 56 years old, and with increased age and large divorce settlements, 53% will be women.

155,500

Until they closed in 1995, Roullet et Decamps were makers of automatons for over 120 years, and the manufacturers of the most expensive toy ever sold – a musical snake charmer – for £155,500.

165,000

Such is the desirability and rarity of truffles that a 3lbs (1.5kg) specimen made nine times its weight when Chinese entrepreneur Stanley Ho paid £165,000 pounds at an international charity auction in November 2007 – outbidding artist Damien Hirst. In 2006, $160,406 was thought to be the highest price ever paid for a truffle – a 3.3lbs (1.5kg) White Alba from Italy. In 2004, a London restaurant bought a similar monster 1.8lbs (850g) truffle for $50,000, and then, unfortunately, let it rot!

1,000,000

Euros paid by a Russian businessman for the ultimate luxury mobile phone, 'Le Million Pièce Unique', by Swiss company Goldvish. A limited edition of only three, this phone is made of 18-carat white gold and encrusted in 120 carats' worth of diamonds.

3,000,000

China clay, or kaolin, has been mined in Cornwall for over 260 years. Three million tons of clay are still produced today, its highest ever production. For every ton of clay, another nine tons of rock, sand and gravel waste are extracted. Almost 70% of some mines' output goes into paper production, where it is used to make sheets glossy and smooth. Other uses are fire-proofing, china, paint production and electrical insulation.

5,338,650

Barney Barnato (1852-1897) was one of the original and greatest mining entrepreneurs in the South African diamond and gold boom of the 1870s. When he sold out his share in the Kimberley diamond mine to his arch competitor – Cecil Rhodes' de Beers Company – for over £5 million, this was the largest cheque ever issued at that time.

9,000,000

The sea's saltwater is estimated to contain as many as nine million tons of gold, but sadly, as it is so diluted, it is very impractical to extract. The rate of dilution is 6 parts per trillion, so a cube of seawater (0.62 miles (1km) deep, wide and long) contains 13.2 lbs (6kg) of gold.

10,000,000

In 2007, Guinness launched a TV ad that cost £10 million to make. Filmed in a village in Argentina and using hundreds of villagers, it shows a chain reaction of objects, starting with 6,000 dominoes, leading on to many other items such as paint cans, fridges, cars, and books, until a tower forms like a pint of Guinness.

12,000,000

The most expensive wedding dress in history is thought to have been the $12 million diamond dress for a wedding show by designer Renée Strauss and jeweller Martin Katz, in America in 2006.

123,000,000

Julius Caesar's decree that no wheeled vehicles were allowed by day in the centre of Rome must have been one of the earliest forms of anti-traffic congestion legislation. Two thousand or so years later, congestion charges are considerable income-raising pieces of legislation, with London receiving net revenues of £123 million.

151,700,000

In just one month in 2008, there were 151.7 million credit card transactions to the value of £11.8 billion by UK residents worldwide – who have a total of 68,683,000 cards, relating to just over 58 million accounts.

339,000,000

The world has a great thirst for champagne and consumed 339 million bottles in 2007 – an increase of 5% over the previous year. Champagne is produced exclusively from 35,000 hectares (83,000 acres) of France around Rheims and Epernay.

400,000,000

After he died in 1991, newspaper and publishing tycoon Robert Maxwell was found to have robbed his companies' pension funds of £400 million, and left 32,000 employees with an uncertain future.

800,000,000

There are over 800 million registered automobiles in the

world today.

950,000,000

Of the estimated 2.7 billion mobile phones in use today, 950 million new ones were sold worldwide in 2007. On average, a mobile phone is replaced every eighteen months.

1,000,000,000

To deal with Zimbabwe's dire economic situation in January 2008, the country issued what was at the time the world's largest value banknote worth Z$10 million – worth about £4, and due to the rampant inflation it became rapidly less valuable. By June 2008, this note had been replaced by a Z$1 billion note, and a special agro-cheque with a face value of Z$50 billion. In September 2007, Zimbabwean inflation was a dramatic 7,982%. However, by March 2008, this had jumped to an even more remarkable 100,000% (a pint of beer costing Z$800 million, a 13 amp plug Z$1.3 billion, and 2 pages of photocopying Z$269 million). All this in a country with a President who has an Economics degree from a London university.

1,400,000,000

There are 1.4 billion credit cards in circulation worldwide.

1,500,000,000

Although there are at least 1.5 billion TV sets in use around the world, given the fact that several people often watch the same set , the actual viewing audience is considerably greater.

860,000,000,000

£860 billion is the annual collective pre-tax pay packet for the whole of the UK's workforce. Whilst the average wage in the UK is £25,000 (of which £4,400 is paid in tax) 0.1% earn more than £350,000 per annum, of whom about 47,000 have an average pre-tax income of £780,000. The average annual Irish pay packet is €33,800 (£26,700).

4,200,000,000,000

The incredible number of German marks that could be exchanged for one US dollar at the height of the country's inflation in the 1920s.

0

The painting of the *Mona Lisa* shows her with no eyebrows or eyelashes; however recent research shows that at some stage Leonardo De Vinci had painted her with eyebrows.

0

At no time did the literary detective, Sherlock Holmes, say his supposedly famous phrase, "Elementary, my dear Watson". In fact this phrase owes its origins to Basil Rathbone's 1930s films. In Conan Doyle's book, *The Crooked Man*, there is one exchange between Dr. Watson and Holmes when, to the doctor's amazement, the detective finishes the logic of how he deduces that the doctor had being doing his rounds by Hansom carriage, by saying "Elementary". He also says "It was very superficial", and "Exactly, my dear Watson" in three other stories. But never "Elementary, my dear Watson".

0

None of Leonardo da Vinci's (1452-1519) paintings are signed, thus attribution can be a matter of dispute. But there are 15 significant works of art (panels and paintings) definitely acknowledged to be by the artist. There are 7 paintings in existence confirmed as by the master, and others attributed – 4 in Italy, 4 in France and 1 each in London, Munich and Russia. Two others are acknowledged to have been painted in collaboration with other artists; a further 6 are possibly by Leonardo, and another 6 are disputed, but at some time have been thought to have been his work.

0

The Celtic languages (Irish and Scottish Gaelic and Breton) have no actual single words for either 'yes' or 'no', but it is quite easy to give an affirmative or negative by using an appropriate verb form in reply to a question.

0

When William Shakespeare created Shylock in *The Merchant of Venice*, it is probable that he had never ever met a practising Jew.

They were banned from England between 1290 and 1566. However, it is possible that he may have met some of those who had converted to Christianity and stayed in the country.

0

...similarly, there is no evidence that William Shakespeare ever travelled outside England, including Scotland, despite the fact that he wrote *Macbeth*. He wrote about Italy in *Two Gentlemen of Verona*, *Romeo and Juliet*, and *The Merchant of Venice*, but never had firsthand knowledge of the country.

0

The mark that the famous French novelist Emile Zola (1840-1902) was awarded for literature at the Lycée St Louis.

0

The amount of glue on an Israeli postage stamp that is not kosher.

1/10

... and 1/10th of a calorie is ingested from licking a stamp.

1

One dead language revived from the past is modern Hebrew, which became the common language of the State of Israel.

1

Of his 872 paintings, 1,019 drawings and 150 water colours, artist Vincent van Gogh (1853-1890) only ever sold one during his lifetime – *The Red Vineyard at Arles* for FF400. During the 10 years that he dedicated to painting, Van Gogh was supported by his brother Theo to the tune of FF150 a month. He was 27 years old when he first took up drawing and painting.

1

One Armed Bandit – coin-operated gambling machine that was originally operated by pulling a lever on one of its sides (hence the one arm), to produce a combination of numbers or symbols that will win...or not (hence the 'bandit')! The payout varies, but is around 95% – thus for every £10 put in only £9.50 will be returned! The first machines were developed in the USA in the 1890s; they are now operated by pushing a button to activate a random selecting computer programme.

1

'One' features in quite a number of well-known sayings or phrases, including 'one for luck', 'one for the pot', 'one for the road' (a now somewhat unacceptable one last drink), 'one red rose' (a simple expression of love), 'one degree under' (not feeling very well or hungover – possibly because of that one drink for the road!), and 'one-on-one'.

1

Only one word in the English language begins and ends with 'und' - underground.

2

The Two Towers, the second volume of J.R.R. Tolkien's *Lord of the Rings* series, and subsequently made into a blockbuster film in 2002 which won two Oscars.

2

There are only two words in English that contain all the vowels in their correct order – 'abstemious' and 'facetious'.

2

Two is another number that features quite often in sayings such as 'two sides of a coin' (acknowledging that there might be two points of view in a argument), 'two-faced' (meaning that somebody is not completely honest) and 'two-timing' (cheating in love).

3

The Three Graces – from Greek mythology – were the three daughters of the god Zeus and the nymph Eurynome, and the goddesses of Joy, Charm and Beauty. Perhaps best known through the Canova marble statue finished in 1814.

3

In Greek mythology, the gates of Hades were guarded by the three-headed hell hound Cerberus who also had a snake for a tail. The other guard dog, that ensured that only the dead could enter and none could leave, was his two-headed brother Orthus.

3

Ménage à trois – a living arrangement comprising three people, not necessarily all the same sex, implying a sexual liaison between some or all of them. Not to be confused with a threesome, which can also be a sexual relationship, but one of a much more temporary and short-lived nature.

3

The number of letters of the word that defines the shortest unit of time – 'now'!

3

There are three basic degrees of Freemason – Entered Apprentice, Fellow Craft and Master Mason, controlled by self-governing Grand Lodges. Beyond Grand Lodges, there are additional Masonic Orders such as the Scottish Rite, with degrees numbered 4 to 33. The first record of an English initiation was in 1646, with the first English Grand Lodge organised in 1717, followed by those for Ireland in 1725 and Scotland in 1736.

3

Three bears in the traditional Goldilocks children's story, which was first recounted in 1837 by the poet Robert Southey.

3

Three exclamation marks are used in newspaper chess columns to indicate an outstanding move. The earliest known chess column appeared in *The Lancet* medical magazine in 1823.

3

The Three Musketeers – written by Alexandre Dumas in 1844 – featuring Aramis, Athos and Porthos, quickly joined by the youthful D'Artagnan. These four heroes in the book were renowned for their swordplay rather than using their muskets.

3.3

Oscar Wilde's prison number was C.3.3 in Reading, where in 1895 he spent much of his two year sentence after being found guilty in the famous trial for indecency. His time in prison inspired two of his great works, *De Profundis* (a 50,000 word letter to Lord Alfred Douglas) and the poem, *The Ballad of Reading Gaol*.

4

Artist and sculptor Michelangelo (1475-1564) took four years (1508-1512) to paint the ceiling of the Sistine Chapel, which includes at least 300 figures. Despite popular belief that he did this lying on his back, in fact he painted standing up on scaffolding of his own design.

4

The Q'uran permits a Muslim man to have four wives. Furthermore, it stipulates that a man must be responsible for the maintenance of each wife. If he does have more than one wife, then he has to provide separate accommodation for each.

4

Number of Presidents' images carved into Mount Rushmore – those of George Washington, Thomas Jefferson, Theodore Roosevelt and Abraham Lincoln. They were carved between 1927 and 1941 by Gustzon Borglum and 400 workers at the relatively modest cost of $1 million. 800 million pounds of stone were removed to create the 60 feet statues (with 20 feet long noses and 18 feet wide mouths).

4

Originals of the *Magna Carta* (1215) in existence – two are held by The British Library, and one each by Lincoln and Salisbury Cathedrals. This first version only protected the rights of the nobles, but it was revised during the 13th century to encompass all citizens, and the final version was made law by Edward 1 in 1297. There are thought to be 13 original copies of this final edition still in existence.

4

Until 1983 the US State of Alaska covered four time zones, after which Yukon, Alaska and Bering were all combined, and the Aleutian islands co-ordinated with Hawaii.

Five-0

The American TV police series *Hawaii Five-0* ran for 12 years from 1968 to 1980. The fictional Hawaiian police department's name was probably based on the fact that Hawaii is the USA's fiftieth state. The lead, with an impressively immovable hairstyle, was

Steve McGarrett, played by Jack Lord.

5

A 'limerick' is a 5-line verse where lines 1, 2 and 3 all rhyme and are written with three metrical feet, while lines 3 and 4 rhyme with each other and are 2 metrical feet. The origin of the name 'limerick' is obscure, but Edward Lear (1812-1888) made the modern form popular. However, the earliest recorded poem in this format was written by Thomas Aquinas (1225-1278).

5

Grenadine is the often forgotten ingredient of a genuine Buck's Fizz, and five drops should be added to the other ingredients – champagne and freshly squeezed orange juice. The cocktail was invented in 1921 by the barman at London's Buck's Club, and is known in the USA as a Mimosa.

5

The Five Pillars of Islam are the basic tenets of Sunni Islamism, and comprise the profession of faith, regular prayer, alms tax, fasting during Ramadan and the Hajj (pilgrimage to Mecca).

5

Number of cattle to every two people in Argentina.

5

Since 2002 the French President holds his office for 5 years – prior to then it was 7 years. While in office he is also ex-officio co-Prince of Andorra. The US President is limited to a maximum of two 4-year terms in office, and elections for the office of President of Russia (instigated in 1991) are every 4 years. In Germany, the President, whose role is arguably more constitutional, is elected every 4 years by a specially convened body – the Federal Convention.

5th

Fifth Column – a clandestine group prepared to undermine a country's solidarity. A term originally coined in the Spanish Civil War (1936–39) when a Nationalist General was marching toward Republican-held Madrid with four columns of troops, and was expecting support from a 'fifth column' of civilian supporters in Madrid.

5

The Famous Five – children's adventure stories by the prolific Enid Blyton (over 600 books). The heroes are brothers and sister Dick, Julian and Anne, and their cousin George (tomboy Georgina), not forgetting the fifth – Timmy the dog! Enid Byton also wrote *The Secret Seven* series.

6

Of the 7 original Gaelic languages, 6 are still spoken today to some extent. It is estimated that in Brittany over 530,000 speak Breton as a first language; in Ireland 355,000, in Wales 575,100 speak Welsh (including 32,700 monolinguists), and in Scotland 62,000 speak Gaelic. Very few speak Cornish or Manx (although there are attempts to revive these two languages). Hiberno-Scottish (based on 12th century Irish) is now totally extinct, although it was used in literature until the 18th century.

6

Regarded by the Greeks as the perfect number, because it is the sum of all its divisors except itself.

6

The maximum number of times you can fold a £5 note by hand.

6

There are six categories of Nobel Prizes awarded each year – started in 1901 as a result of the will of the Swedish inventor of dynamite, Alfred Nobel (1833-1896). Five of the prizes – Chemistry, Economics, Medicine, Literature and Physics – are awarded by Swedish committees. However, as specified in Nobel's will (but with no explanation) the Prize for Peace is awarded by a Committee from the Norwegian Parliament. There have been 777 individual Nobel Laureates (only 34 of whom have been women) and 20 winning organisations (The Red Cross being a winner on three occasions).

6

The original name for the Sistine Chapel was in fact 'Sixtine', after Pope Sixtus IV (1414-1484) who inspired its construction in the late 15th century.

7

T.E. Lawrence's original book *The Seven Pillars of Wisdom* was going to be a scholarly book about seven great cities of the Arab world. Once WWI started, he destroyed this manuscript, and then later used the title for his autobiography of his Arabian exploits. While changing trains at Reading in 1919, he lost the only draft of the manuscript, and then had to rewrite the whole book again from memory. The first edition (of 200 copies, with hand-crafted binding) is one of the most valuable antiquarian books in existence.

7

When the 'Flood' came, as reported in the book of Genesis in the Bible, Noah was instructed by God to take in two of 'every living thing'. But further on, there is another confusing instruction '...of every clean beast thou shall take to thee by sevens, the male and his female, and beast that are not clean by two...'.

7

Ages of man, according to Shakespeare in the play *As you Like It*, that opens with the famous lines 'All the world's a stage...' and then describes the ages as the infant, the whining schoolboy, the lover, the soldier, the round-bellied justice, the sixth stage as being lean and slippered, and finally, second childishness – 'sans teeth, sans eyes, sans taste, sans everything'.

7

There are seven 'Cardinal Virtues' – categorised in two sections. First, there are the 'theological': faith, hope, chastity, and second the 'natural': justice, prudence, temperance, fortitude. These together with the seven 'Deadly Sins', were designated by Pope Gregory in AD60.

7

...and the seven 'Deadly Sins' are anger, covetousness, envy, gluttony, lust, pride and sloth,

7

In ancient Irish tradition the seventh son of a seventh son has magical powers. However, in Romanian folklore he would be a vampire. The heavy metal rock band, 'Iron Maiden', issued an album called *The Seventh Son of the Seventh Son* in 1988.

7

In Italy, Spain, Mexico, Brazil and the Middle East, superstition dictates that a cat has seven lives. However, their feline counterparts in Britain, USA, Russia and China are luckier and credited with 9 lives.

7

Seven is thought to be a 'lucky number' in some cultures. In the West, this is possibly because the Bible states it took the Lord six days to create the world, and that on the seventh day he rested; or perhaps because the opposite sides of dice each add up to 7. On a Tarot card the seven of clubs indicates 'ambition, drive, and desire'. In Chinese culture, the seventh day of the first moon of the lunar year is known as Human's Day, to be celebrated as the universal birthday of all human beings.

7

The number 'seven' has a significance in quite a number of religions – in Judaism, God sanctified the seventh day, a sabbatical comes every 7 years, and there are 7 blessings in a Jewish wedding. In Christianity there are not only seven virtues; there are the same number of deadly sins, and also the 7 Joys of the Virgin Mary. In Islam there are 7 Heavens as well as Earths, and there are 7 doors to both heaven and hell. Seven is also a very important number in the Cherokee Indian tradition; in Buddhism, Buddha walked 7 steps at his birth, and in Hinduism there are Seven Promises and Seven Rounds in a wedding ceremony.

7

Wonders of the World – Great Pyramid at Giza (2650-2500 BC) – still standing; Hanging Gardens of Babylon (660BC), destroyed by earthquake 1st century BC.; Temple of Artemis at Ephesus (550BC), destroyed by arson 356BC; Statue of Zeus at Olympia (435BC), destroyed by fire (5-6th C BC); Mausoleum of Mausollos at Halicannassus (351BC), destroyed by earthquake 1404 AD; Colossus of Rhodes (292-280BC), destroyed by earthquake 224 BC; Lighthouse at Alexandria (3rd C BC), destroyed by earthquake AD1303-1400.

7.2

Judged by her head size, if the Barbie doll were lifesize, she'd be 7.2 feet (2.2m) tall, and strangely, given she is meant to be a teenage girl, her figure would be 36-18-38! Introduced in 1959, her full name is Barbie Millicent Roberts, and somewhere in the world two Barbie dolls are sold every second. She has over 80 careers, from teenage fashion icon and rock star to army medic, and from Olympic swimmer to Presidential candidate.

8

Although the ages of the world's youngest parents are thought to be 8 and 9 in China in 1910, the youngest ever mother is Peruvian Lina Medina, who, at the age of 5 years and seven months, gave birth to a baby boy in 1933.

8

Bells – system of bells on board a ship to regulate watches. The strikes of the bell do not accord to the number of the hour. Instead, there are eight bells, one for each half-hour of a four-hour watch.

8

The number of French opponents that Paul Charles Morphy (1837-1884), considered by many the greatest chess player in the world, aged 22, played at once – blindfold. Memorising the changing position of 256 chess pieces, he won 6 of the games, tied one and lost one.

8

Number of records selected by each guest celebrity to take to a mythical desert island in the Radio 4 programme 'Desert Island Discs'. The programme started in 1942, and the most requested record is *Ode to Joy* from Beethoven's 9th Symphony.

8 to 1

Proportion of gin to vermouth for the perfect very dry martini (however proportions may vary according to taste). H.L. Mencken said that this cocktail was "the only American invention as perfect as the sonnet".

8.5

The Oberammergau Passion Play depicting the life and death of Christ takes 8.5 hours including intervals. It has been performed every 10 years since 1634, when the inhabitants of the Bavarian village were spared from the bubonic plague. Over 2,000 villagers make up the cast, singers, instrumentalists and stage hands.

9

Normally a Jewish candlestick – 'menorah' – has seven branches. However, the Hanukkah Menorah has nine – eight of them for each day of the Hanukkah holiday, which commemorates the rededication of the Temple. To celebrate, a candle was lit with enough oil for one day, but remarkably it lasted eight days.

10

The word Œtherein, contains ten words without rearranging the letters; the, there, he, in, her, here, herein, therein, rein and ere.

10

The 17th century poet John Milton (1608-1674) sold the rights to his best known-poem, *Paradise Lost*, for £10 in 1657. This poem is all the more remarkable because Milton was blind when he wrote it.

10

Times the image of the Virgin Mary has appeared on the front cover of *Time* magazine.

10

The first ever speed limit for self-propelled vehicles, introduced in 1861, was 10mph. However this was reduced in 1865 to 4mph in the country and 2 mph in towns, with a man required to walk 60 yards (50 m) ahead with a flag or a lantern to avoid frightening horses.

12

Average number of pairs of shoes a ballerina uses each week.

12

Days of Christmas – starting on Christmas day and finishing on

5 January, the eve of Epiphany. In the traditional song each day is represented by the following – 1 partridge in a pear tree, 2 turtle doves, 3 French hens, 4 calling birds, 5 golden rings, 6 geese a-laying, 7 swans a-swimming, 8 maids a-milking, 9 ladies dancing, 10 lords a-leaping, 11 pipers piping and 12 drummers drumming.

12

The age of consent for marriage for a girl in Britain was only 12 before 1929; for a boy it was 14. Nowadays, the age is 16, but parental (or a guardian's) written consent is required if under the age of 18. In France, a girl could get married at the age of 15 until 2006, when the age limit was raised to 18 to alleviate the possibility of marital violence.

12

To become an authentic and traditional Japanese sushi chef can take up to 12 years, and includes the art of cooking rice, 63 species of saltwater fish, 8 types of shellfish and 18 freshwater fish, as well as learning the associated rituals.

12

Although Jesus may have had many followers during his life, his closest supporters were the Twelve Disciples – Simon (called Peter), Andrew, James (son of Zebedee), John, Philip, Bartholomew, Matthew, Thomas, James (son of Alphaeus), Simon (the Zealot), Jude Thaddaeus and Judas Iscariot.

12

The number of Jane Austen's books – *Sense & Sensibility* (1811); *Pride & Predjudice* (1813); *Mansfield Park* (1814); *Emma* (1816); *Persuasion* (1818); *Northanger Abbey* (1818); and published after her death, *The Watsons and Lady Susan* (1871); *Love & Friendship* (1922); *Collected Letters* (1932/52); *Volume The First* (1933) and *Volume The Third* (1951).

13

A Jewish boy celebrates his Bar Mitzvah at the age of 13, at which time he 'comes of age' and takes on the religious responsibilities of an adult. Traditionally, girls celebrated their Bar Mitzvah at 12, but this can now be 13. If a man reaches the age of 83 he sometimes has a second Bar Mitzvah – signifying him reaching his natural lifespan of three score years and ten (70), plus the 13 up to his first Bar Mitzvah.

13

'Thirteen o'clock' strikes when Big Brother watches over us – a time when man has no right to speak, and no freedom to do anything Big Brother doesn't allow – in George Orwell's classic book *'1984'*

13

It is Tuesday the 13th, as opposed to Friday the13th, that is considered unlucky in the Middle East. Thirteen is accepted fairly universally as being an unlucky number, perhaps because in the Christian tradition, Jesus' betrayer, Judas Iscariot, was the thirteenth to sit down at the Last Supper. Some buildings don't have a 13th floor – the numbering going from 12 to 14, or even sometimes having the number 12B

14

Although the average alcoholic content of wine is between 8 and 10%, it can still be classified as 'table wine' up to 14%. Anything above that will have had additional alcohol added, and be classified as 'dessert' or 'fortified' wine. Sparkling wine normally has 8 to 12 %, and beer only 5% alcohol.

15

Each backgammon player starts with 15 pieces (which can also be known as counters, checkers, chips, men or stones). Each side of the board has 12 triangles, and players move their pieces in counter horseshoe directions according to the results of the two dice they throw alternatively. Backgammon originated in Mesopotamia in 3,000 BC, but the most modern addition to the game was in New York in 1926 – the doubling dice which allows players the extra dimension of increasing the stakes by gambling on their probability of winning.

15

Miles – the length you would have to walk in St Petersburg's Hermitage Museum to visit all of its 3 million works of art in no less than 322 separate galleries.

15

Percentage of French wines labelled '*Appellation Controlé*' – there are over 300 AOC specified wines and 19 spirits.

15

The exposure necessary for a Daguerreotype plate to receive an image was 15 minutes for this early form of photograph. People had their heads clamped in position to stop them from moving. This method was perfected by French chemist Louis Daguerre in 1839 and was not the earliest type of photography, although it allowed the image to be developed considerably faster than its predecessors.

16.38

June 21 – the Summer Solstice, is the longest day in the Northern Hemisphere. London has 16 hours and 38 minutes of daylight, with sunrise at 4.42 a.m. and sunset at 9.20 p.m. On the same date in Reykjavik, Iceland, just below the Arctic Circle, the hours of daylight are much longer – running from 2.58 a.m. to midnight. In Nairobi, Kenya, the day is exactly 12 hours long with sunrise at 6.33 a.m. and sunset at 6.33 p.m.

17

A record-breaking 17 curtain calls were taken by Luciano Pavarotti (1935-2007) after the tenor's performance in Donizetti's *La Fille de Régiment* at New York's Metropolitan Opera in 1972. This was nearly equalled by diva Maria Callas (1923-1977) on her debut in Bellini's *Norma* at the same location in 1956 – she received 16 curtain calls.

21

In his will, Joseph Pulitzer (1847-1911), the pioneering American journalist and newspaper proprietor, left $2,000,000 partly to establish a journalism school and to create prizes for excellence in journalism, arts, authorship and letters. First awarded in 1917, there are now 21 categories of Pulitzer Prize which, although it brings an award of $10,000, is also much valued for the accolade.

21

Total spots on a six-sided dice, with numbers from 1 to 6.

22

Catch 22 – Joseph Heller's 1961 anti-war novel about a US bomber squadron that describes the illogical situation and trap in which the hero tries to get himself grounded by being pronounced insane, but is told that only an insane person would want to fly, and his desire not to fly proves that he is, in fact, sane, and therefore must continue to do so.

23

If 23 people gather together in the same room, then mathematical probability decrees that there is a better than evens chance of two having the same birthday (though not necessarily the same year of birth).

25

It is difficult to know exactly how much William Shakespeare (1564-1616) earned from his endeavours, but it is thought that his annual income was around £25. However, at the height of his fame this might have risen to as much as £200 – some of which he invested in property around Stratford-on-Avon.

27

The Bible's New Testament consists of 27 books written by various authors between 45 AD and 140 AD. There are four books of Gospels, The Book of Acts (of the Apostles), twenty-one Epistles or letters (thirteen of which were written by Paul), and the final book is Revelations. The number of books in the Old Testament varies according to religion – the Protestant Church recognises thirty-nine, while the Catholics count forty-six or forty-seven because they include the extra books of Apocrypha (Greek for 'hidden away').

28

The surprisingly young age at which cookery guru Mrs Beeton (1836-1865) died was 28. Her *Book of Household Management* made a major impact, as it not only provided over 900 recipes but also guided Victorian ladies and their following generations on the complete management of their houses. It is thought that many of her recipes were plagiarised, and not even personally tested, as one of the earliest for a 'Victoria sponge' omitted eggs – a rather essential ingredient.

29

The number of letters in the longest real word in the English language – floccinaucinihilipilification – the meaning of which is 'the act of estimating as worthless'.

36

'There are not 36 solutions', (*'Il n'y a pas trente-six solutions'*), a well-known French phrase indicating that you should make your mind up more quickly.

37

There are 37 numbers on a British roulette wheel including 0. The calculation of roulette odds for the gambler prove that the 'House' has a 27% edge - meaning for every £1 you put on your likely average return will be only 27 pence. This may have prompted Albert Einstein to reputedly say "You cannot beat a roulette table unless you steal money from it". But many, including mathematicians and engineers, have tried by using their own scientific methods or carefully concealed computers to hopefully predict on which number the ball will fall. Few have succeeded. In the summer of 1891, at the Monte Carlo Casino, Charlie Wells, a part-time swindler and petty crook from London, broke the bank at each table he played over a period of several days by winning all the available money in the table bank that day, and a black cloth would be placed over the table until the bank was replenished. In song and life he was celebrated as 'The Man That Broke the Bank at Monte Carlo'. Wells later admitted that it was all luck.

37

The maximum points that can be held in any one bridge hand by an individual with their thirteen cards (counting an Ace as worth 4 points, a King as 3, a Queen as 2 and the Jack as 1).

39

The Thirty Nine Steps is the classic spy story written by John Buchan in 1915, and subsequently made into a highly successful 1939 Alfred Hitchcock film – remade less successfully in 1985.

40

An integral part of the Harrow School year is the 'School Songs' –

attended by pupils and old boys alike. '*Forty years on*' is probably the most famous of these songs; it was written in 1872 but had an additional verse added in 1964 to celebrate Old Harrovian Sir Winston Churchill's ninetieth birthday.

41

41 to 45 degrees Fahrenheit (5 –7 Celsius), the temperature at which it is recommended that champagne should be served.

42

In the Japanese language when the words '4' and '2' are pronounced together they mean death, and thus the number 42 is considered bad luck.

42

The perfect number as defined in *The Hitchhiker's Guide To The Galaxy* by Douglas Adams.

47

Henri Matisse's painting *Le Bateau* hung upside down in New York's Museum of Modern Art for 47 days before any of 116,000 visitors who looked at it noticed.

50

Of the 250 alphabets in the history of language, only 50 have survived. They vary considerably in length, with for example, the Hawaiian alphabet having only 12 letters – a, e, i, h, k, l, m, n, o, p, u, w. (Hawaiian was originally an oral tradition, but it was the 19th century missionaries who put together the written alphabet). There are 24 letters in the Greek alphabet, while the Russian alphabet has 31 characters.

50

The number of uses of a brick – as defined by a well-known intelligence test – vary considerably, with some being very obscure or unmentionable. Other uses include a knife blunter, part of a Turner art prize entry, a puzzle for the Lost Property Office, for throwing through windows, and, not surprisingly perhaps, for building.

55

The world-famous fictional detective and amateur violinist, Sherlock Holmes, bought his Stradivarius violin for 55 shillings (£2.15). Nowadays, when one of the master musical instrument maker's remaining 600 or so creations comes on to the market, the price tends to be millions of pounds.

69

Some of the most prized and romantic antiques in the world are the 69 ornamental eggs made by the Russian jeweller family of Fabergé. Thanks to the ravages of the revolution, only 61 remain today. The court jewellers' main clients were Tsars Alexander III and Nicholas II, the last Tsar of Russia, for whom they made 50. The 'Rothschild' egg (made in 1902) was sold at auction in November 2007 for £8.9 million.

70

Three score years and ten, the Biblical age of an old man.

100

An average size (150 ml) cup of coffee contains 100 mg of caffeine. In comparison, a cup of tea has 39 mg, a cola or similar caffeinated soft drink 15 mg, and both hot chocolate and decaffeinated coffee only 2 mg.

100

The slang term, the 'ton', is the speed of one hundred miles per hour (mph). The precise origin is unknown, but it came into common usage in the 1940s and 50s when motorbikers used get to get together and it became a challenge to reach 100 mph.

101

The room in George Orwell's *1984*, to which Big Brother could send you and where you would find what you feared the most. Thought to originate from a conference room at the BBC1. Ironically, *Room 101* became the title of a BBC TV series in 1994 where celebrity guests selected people, things or places that they thought the world would be better without – thus consigning them to Room 101.

119

With 176 verses, divided into twenty-two stanzas of eight lines each, Psalm 119 is the longest in the Bible. It is one of a number of acrostic poems in the Bible, with each stanza starting with successive letters of the Hebrew alphabet.

120

120 Days of Sodom by the Marquis de Sade was written in 1785 on a 39 feet (12m) roll of paper in tiny writing, as he was short of materials while imprisoned in the Bastille. The book was not officially published until 1905 in France, and was subsequently made into an Italian film in 1975.

168

The total number of dots on the 28 tiles that make up a set of dominoes.

200

Although variations of the bikini have long existed, (some of the earliest images are on Greek urns dated 1500 BC) the invention of the modern version is attributed to French engineer Louis Réard whose 1946 miniature swimsuit was only 200 square inches (1,290 sq cm). Its name comes from the Pacific atoll where nuclear bombs were tested, as it was thought that its introduction would cause a similar sensation. Certainly three countries, Italy, Portugal and Spain, initially banned it as being indecent. One contemporary comment was "the bikini reveals everything about a girl except her mother's maiden name".

253

The London Underground, the Tube, consists of a network of 253 miles. When the first section between Farringdon Street and Paddington opened in 1863, it carried 30,000 passengers a day. Today's 12 lines carry an average of 3.5 million passengers and serve 287 stations (only 29 of which are south of the River Thames) with a total of 412 escalators (the longest of which is 197 feet at the Angel). However, the longest route is the Central Line, running for 34 miles from West Ruislip to Epping ,and the shortest is Waterloo & City (the 'Drain'} with only two stations on its 1.2 miles.

364

This is the total of the face points in a pack (or deck) of 52 cards – not only counting the number cards, but also the court cards, with the following values – Jack 11, Queen 12, and King 13. However, the Ace can be either the least value – 1 (as we have calculated here) – or the highest, worth 14. The pack of cards, as we know it today, with the four suits (spades, hearts, diamonds and clubs) probably originated in France in the 1480s, having been adapted from earlier variations.

365

To celebrate an epic moment in the American Civil War, a number of entrepreneurs commissioned a 'cyclorama' (360° cylndrical painting) of The Battle of Gettysburg by Paul Phillipoteaux. When it first went on show in 1883, to add realism it incorporated actual piles of earth, walls and artefacts from the 1863 battle. Various other versions were created, but the original was thought to be 365 feet (111 m) in circumference and 42 feet (12.8 m) high. As a 'hidden' signature the artist included himself in the picture wielding a sword.

400

In 1911, British Members of Parliament were paid for the first time ever. A salary of £400 a year was introduced, with Lloyd George stating that this was the minimum sum needed "to attract men who would render incalculable service to the state". This has risen to a basic sum of £61,820, plus considerable, and currently controversial, allowances for expenses.

451

Fahrenheit 45, the temperature at which books will burn – in Ray Bradbury's 1953 futuristic book. He depicts a world without books, and one in which firemen start fires rather than put them out. Made into a film in 1966.

464

The Oxford English Dictionary lists 464 definitions of the word 'set'. The next most numerous definitions for a word is 396 for 'run'.

550

Oil paintings completed by artist Joseph Mallord William Turner (1775-1851). His prodigious output also included over 2,000 watercolours and 30,000 drawings and sketches on paper.

613

The number of Commandments in the Jewish religion – including the 10 in the Christian religion.

666

From the Bible known as the Mark of the Beast – from Revelations 13:18. "Here is wisdom. Let him that hath understanding count the number of the beast: for it is the number of a man; and his number is six hundred threescore and six". One interpretation is that, when encoded, the number represents the Antichrist.

911

The emergency number in the USA – in the UK this has traditionally has been 999 as this was the easiest configuration to find in the dark on an old finger-dialling telephone.

1001

The origins of *1001 Arabian Nights* lie in Arabian writings of around 850 AD. The key of the story is how King Shahryar was convinced by his young virginal bride Scheherazade not to behead her, the fate of all his other brides after one night, if she could amuse him with her story telling. This she must have done for 2 years and 271 nights (total of 1001). Her stories include such classics as *Aladdin*, *Sinbad the Sailor* and *Ali Baba and the Forty Thieves*. Sir Richard Burton was thought to be the main European translator in 1850. In fact, Frenchman Antoine Galland was the first, between 1704 and 1707.

1,327

In chess, there are 1,327 recorded opening moves and variants. However, there are a further 318,979,564,000 possible permutations when playing the first chess four moves on each side.

1,700

Completely new words created by William Shakespeare amongst the 17,670 words used in his plays. He also created many phrases never used before; 'in one fell swoop', 'in a pickle', 'play fast and loose', 'cold comfort', 'foul play', 'tower of strength', 'to the manner born', 'a sea of troubles' and 'milk of human kindness'. His works also incorporated 10,000 puns.

1,775

The number of poems written by American poet Emily Dickinson (1830-1886) – nearly 800 of which were discovered after her death. Her poetry is all the more extraordinary as in the last fifteen or so years of her life she became very introverted and reclusive and hardly ever left her house – sometimes talking to visitors from behind closed doors.

1984

The title of George Orwell's ninth book, written in 1949, predicting with uncanny accuracy a highly centralised and controlled world.

2,000

J.K. Rowling, the author, was offered £2,000 as the advance for her first Harry Potter book. She has subsequently gone on to earn £550 million from further book deals and films. The profits from the overall brand, including films, DVDs, video games and merchandising, is estimated to be £10 billion.

6,000

When Charles Dickens wrote *A Christmas Carol* in 1843, he thought it would merely help him pay off a debt. In fact, it sold an amazing 6,000 copies in the first week. He had paid for the publication himself, but because he made it quite a lavish production and sold it for only 5/- (5 shillings = 25p today), he didn't make that much money to start with.

6,912

Languages in the world. However, 50% will disappear before the end of the century. The UN estimates that one dies every two weeks. 20 to 40% of the languages are already moribund, while only 5% are widely spoken or are designated as official languages.

9,000

In keeping with his traditional position of absolute power as the King of Siam – Mongut (1809-1868) – had 9,000 courtesans. He also had 82 children by 39 wives. He was made famous when British army officer's widow Anna Leonowens was appointed governess to his children, and subsequently wrote two books of her experiences, which in turn were made into the long-running musical '*The King and I*' starring Yul Brynner, and a 1999 film '*Anna and the King*' with Jodie Foster.

10,000

The international French organisation to promote the wines of Burgundy, the '*Confrérie des Chevaliers du Tastevin*', has an international membership in the region of 10,000. Although the brotherhood was originally formed in 1703 under a different name, its present form originates from 1934 and is based on the Château Clos de Vougeot.

11,000

In Venice of the 1600s and 1700s, courtesans thrived and the city became famous for them. It was reckoned there were as many as 11,000, contrasting only 800 nobles' wives. They were so obvious, and in many cases, dressed as well the noble ladies, that they were banned from wearing similar clothes and jewellery.

13,500

Pablo Picasso (1881-1973) was one of the most prolific artists of all time, and by his death in 1973 he had produced 13,500 paintings and designs, together with 300 sculptures and numerous ceramics and prints. At that time the collection was worth 1,251,673,200 francs – ($250 million or £102 million) - but has risen vastly in value over the years.

14,500

Episodes of the radio soap *The Archers* since 1950, when it started as an agricultural advice programme to ease the effects of food rationing.

20,000

Leagues Under The Sea – the 1873 novel by French pioneer-science fiction author Jules Verne – featuring submarines long before they were a reality. In the same way, Verne wrote *Around the World in 80 Days* at a time when this was an almost unimaginable feat. Both were made into films.

64,000

'Sixty-four thousand dollar question'. This common phrase in Britain to mean 'the most important question' originates from an American game show of the same name, although, strangely never actually broadcast in the UK. Contestants were asked questions on their chosen subjects, with the initial prize being $1 for a correct answer, and then each subsequent prize doubled up with a correct answer until the 17th was worth $64,000. Despite its immense popularity, the show only lasted from 1955 to 1958.

65,000

In 1865, the circulation of *The Times* was 65,000 – the most widely read newspaper in the world at that time.

98,721

The extraordinary number of letters that Lewis Carroll wrote in the last 37 years of his life – averaging 7 a day.

100,000

The 100,000 francs that Claude Monet (1840-1926), artist and founding father of French Impressionism, won in 1891 on the French National Lottery. This helped to sustain him during a precarious financial lifetime of painting.

117,000

Books that Abdul Kassem Ismael, the Grand Vizier of Persia, carried wherever he went, in alphabetical order, on 400 camels.

144,000

Based on the Bible (Revelations 7:4-8) only 144,000 Jehovah's Witnesses will go to heaven to rule with Jehovah God. Anybody born after 1936 cannot achieve this status and awaits earthly resurrection in the Millennium. This religion (also known as the Watchtower Bible and Tract Society) was founded by Charles Taze Russell in Pennsylvania, USA in 1884.

200,000

The book *The Diving Bell and the Butterfly (Le Scaphandre et le Papillon)* is a remarkable memoir written by French journalist Jean-Dominque Bauby after he had a stroke. His only method of communication was by blinking his left eye in a code for each letter. Words took about two minutes, and over 200,000 blinks later the book was published to critical acclaim in 1997. Two days later, the author died.

400,000

In 900 AD, Cordoba (in modern Spain) was one of the great cultural cities of the world. For example, its library contained a remarkable 400,000 books.

995,112

Trying to define the number of words in the English language has proved to be an interesting challenge. The Global Language Monitor has come up with the figure 995,112 and includes jargon and scientific terms. However the 2nd edition of the *Oxford English Dictionary* has 171,476 current words (1/2 nouns, 1/4 adjectives and 1/7 verbs) and 47,156 that are obsolete. To highlight the difficulty and the problem of what is included or not, the American Chemical Society have named 84 million chemicals; and if you wrote out every number between 0 and 999,999, that would be one million words alone!

311,875,200

There are over three hundred and eleven million variations of dealing a five card sequence from a standard pack of 52 playing cards, and from which 7,462 unique poker hands can be dealt.

4,320,000,000,000

In the Hindu religion there are 4.32 billion solar days to one day of Brahma (the God of Creation), and the same number of nights.

43,252,003,274,489,000

Total possible combinations of Eeno Rubik's cube puzzle – invented in 1974. When Rubik cube mania first hit the world, over 100 million cubes were sold between1980-82. 26 is the fewest number of moves needed to complete the puzzle. Six of the squares on a cube never alter position.

0.004

Inch – the size of a woman's egg, which is the largest cell in her body. On the other hand, a man's sperm at 0.000025 inches is the smallest cell. About 2,500 sperm in a single layer would cover the full-stop at the end of this sentence.

1

In one breath an adult inhales 1 pint of air, although the actual capacity of a lung is 5 litres (10 pints). The lungs consist of 1,500 miles of airways, and the air-gathering alveoli have the surface area of a tennis court.

1

In one hour, the rate at which a 150 lb (67.5 kg) human body will dispose of alcohol is either 12 ounces (340 g) of beer, 4 ounces (113 g) of wine (1 glass), or 1 ounce (1 shot) of hard liquor.

1

Pound (0.45 kg) of fat lost if you walk for 24 hours. However, by walking an extra 20 minutes each day, an average person would burn off 7 pounds of body fat in a year.

1

The human skin is the largest organ of the body. In each square inch, there are 4 yards of nerve fibres, 1,300 nerve cells, 30 sebaceous (oil) glands, 3 million cells and 3 yards (2.7 m) of blood vessels.

1.5

A crematorium takes one and half hours to consume a body entirely.

1.5

To survive efficiently in a hot desert, a human needs to drink 1.5 gallons (6.8 l) of water each day. Interestingly, very cold dry air can be equally dehydrating. Without water, the human body can lose up to 10% of its weight without any major ill effects. However, once 25% of the body's water has been lost, there is a corresponding 25% loss of efficiency.

2

Most of the vital organs in the body are duplicated as a back up – the ovary, testicle, kidney and the lobe of the liver. Unfortunately, we only have one heart!

2

According to Swedish research, those of the 1960s "make love not war" generation, who are now in their seventies, are enjoying almost twice as much sex as previous generations – this includes both married couples (men up from 52% to 98% and women up from 38% to 56%) and unmarried (men from 30% to 54% and women up from .8% to 12%). However, surprisingly women reported a higher level of sexual satisfaction, while men reported lower satisfaction.

3

Months – the first time a baby can actually shed tears; in fact, humans are the only creatures that cry when upset.

3

The length of the human body is three times the circumference of the head.

3

The vagina in a grown woman is about 3 to 5 inches (7.6-12.7cm) long – but because it has muscular walls it can expand or contract as required. The vagina leads to the uterus, which, when a woman isn't pregnant, is about 3 inches (7.6 cm) long and 2 inches (5 cm) wide. The 4 inch (10.2 cm) long fallopian tubes connect the uterus to the two ovaries, each measuring 1 to 2 inches long.

3

Three per cent of the body's weight is made up by the brain – the average weight being 3.3 lbs (1.5 kg). However, the brain accounts for 20% of our total oxygen consumption.

3

Years – the amount of time that a woman supposedly spends in her lifetime 'getting ready to go out'.

4

In an average lifetime, a man spends four months shaving. A man's beard grows 5.5 inches (14 cm) a year.

4

In the ancient Hippocratic medical tradition, the human body was governed by four 'humours', which in turn related to the seasons of the year – blood (spring), phlegm (winter), yellow bile (summer) and black bile (autumn). The ancient doctor's intent was to achieve a balance between these humours for his patient's wellbeing.

4

Medical opinion states that the human body can survive 4 to 6 weeks without food, but this depends on the individual's initial condition. The very obese can live from 3 weeks up to 25 weeks and more without food, depending on the amount of fat and how much water is taken. In 1981, Irish political prisoners lasted between 46 to 73 days on hunger strike. Mahatma Gandhi (1869-1958) survived a 3-week fast in 1947 when he was in his 70s. However, the body can last only 10 minutes without air.

5

It takes 5 months to grow a new fingernail from base to tip, and fingernails grow four times faster than toenails.

5

The human tongue can taste only five flavours – bitter, sweet, sour, salt and umami (meaty, savoury taste associated with protein-rich foods). The tongue has about 10,000 tastebuds, but the specific receptors are contained in certain areas – sweet at the tip, sour at the sides and bitter with umami at the back. Only the salt tastebuds are spread evenly around the tongue.

6

'Blood is thicker than water' by a factor of 6.

6.5

Human nasal hair grows at the same rate as the rest of body hair, and over a lifetime a nasal hair could theoretically grow up to 6 feet (1.8 m).

7

Arduous exercise can lead to a human body sweating off as much as 7 pints (4l) in a day.

7

Average time (in minutes) it takes to fall asleep.

7

Food takes about 7 seconds to pass from the mouth to the stomach via the oesophagus, and will then spend about 6 hours in the stomach being digested.

8

An adult body contains 10 pints (5.7l) of blood, which accounts for about 8 per cent of our weight.

10

Despite a popular myth that only men are colour blind, 10% of women are also colour blind – the same percentage as men.

12

The average adult human stomach is about 12 inches (30.5 cm) long and 6 inches (15.25 cm) at its widest point. Its capacity is just less than 1.76 pints (1 litre).

15

Blood consists of a number of different cells, the smallest of which are the platelets whose primary function is to begin the process of coagulation by clumping together in the area of the injury or cut. They also emit messages into the blood to attract more platelets to plug the wound, and to get the plasma protein, fibrinogen, to make a mesh over the wound. This can take up to 15 minutes.

16

It is estimated that as many as 16% of the population are left-handed. Writing with the left hand is the most obvious indicator, but it is difficult to give a precise percentage because for many years pupils were taught to write only with their right hand. As a result, a number display ambidexterity. Amongst famous rock stars, Jimi Hendrix and Paul McCartney originally played their right-handed guitars upside down to accommodate their left-handedness. Beatles drummer Ringo Starr is left-handed, but plays the drums right-handed.

16

Percentage of alcoholics who are women.

16

The average human farts 16 times a day, and the pint of intestinal gas released – some due to inhaled air and the rest from fermentation of undigested food – is, perhaps, an unacknowledged contribution to global warming!

20

About 20% of human body weight derives from the 206 bones that make up the skeleton. Babies are born with about 300 bones, but as they mature a number of these fuse naturally.

24

Weeks – the UK time limit for legal abortions. On 22 May 2008, Parliament voted by 304 votes to 233 not to lower this figure to 22 weeks.

25

The average amount a human blinks per minute, with an average blink taking 300 to 400 milliseconds. An eyelash lasts about 150 days.

27

The fastest speed a human can run is 27 mph (43.4 kph).

28

Each human foot contains 28 bones, and each hand has 26. Since the human body has a total of 206 bones, the two hands and the two feet account for over 50% of the total.

28

The number of sexual positions in the *Kama Sutra* – the 2nd century Sanskrit text that is as much about the acquisition of knowledge and happiness as it is about sexual virtuosity.

28

The outer layer of skin is replaced every 28 days (and the dead skin is estimated to account for 90% of household dust).The skin is about 2 square metres in area, and accounts for 1/8th of the body's weight. It is the first line of defence for the body against unfriendly environments and germs, and interestingly it is waterproof in both directions – but allows sweat!

28

The smallest commercial bra size is 28AA – anything smaller probably negates the need for any bra at all. The largest commercial size is 48DD, although there are reports of an internet seller who offers an amazing 56FF bra. The most substantial recorded 'natural', as opposed to enhanced bosom size is that of a US woman's remarkable 56WW. (The number relates to the circumference of the upper body, and the letters to the cup size).

29

There are a total of 29 internationally recognised blood groups, of which Rhesus Positive is the most common.

32

Adult humans have 32 teeth of 3 different types. The 4 incisors are used for biting, the 2 canines for tearing at food and the remainder are molars which act as grinders. At about 5-6 months, children start to grow 20 milk teeth that last until they are 5 to 6 years old, when they are replaced by adult teeth.

37

The brain continues to send out electrical signals for 37 hours after death.

40

Miles – the length of tubes in a kidney.

40

Winks – the euphemism for having a quick sleep. The origin is uncertain and the phrase strange, as one does not wink when asleep!

40-20-37

Country & Western singing legend Dolly Parton's remarkable vital statistics. She kept them a secret for a long time, until she eventually admitted "I used to never tell anybody my measurements, but the older I get the more proud I am."

46

Chromosomes are thread-like double helix strands of DNA (deoxyribonucleic acid), which carry the genes and transmit heredity information. A human normally has 46 chromosomes (23 pairs) in all but the sex cells. Half of each chromosomal pair is inherited from the mother's egg, and the other half from the father's sperm, and when egg and sperm fertilise, they then create a single cell with 46 chromosomes. Each tightly-packed DNA coil contains about 35,000 genes, and, when unravelled, is about 6 feet 6 inches (2 metres) in length.

50

It is estimated that an average person in the western world eats about 50 tons (50,800 kg) of food in a lifetime – as an adult, this is about 1 ton a year.

50

Per cent of body heat escapes through the head.

50

Seconds – the time it takes for the blood to circulate completely round the body. William Harvey (1578-1657) is credited with being the first man to identify the circulation of blood, and the role played by the heart as the pump. The average lifespan of a blood cell is 120 days.

50

When a girl is aged nine, she is 50% of her optimum adult weight.

52

Daniel Lambert was at one time Britain's fattest man. Aged 39 in 1809, he weighed over 52 stone (336 kg). His height was 5 feet 11 inches (1.8 m), but his waist measurement was 9 feet 4 inches (2.8 m). When he died in a pub, a wall had to be knocked down to remove his body.

55

1 in 55 births in Canada occurs in a car. One wonders how many are conceived in a car!

95

Pounds (43 kg) – the weight of oxygen in the average adult human body – 65% of its weight. The next most common elements are carbon at 35 lbs (15.9 kg), and hydrogen at just over 15 lbs (6.8 kg).

99

Ninety-nine per cent of babies are born without moles, with the remaining 1% having congenital nevi or moles. Most moles appear during the first 20 years of life, although they may continue to develop into the 30s and 40s.

99.99

Per cent of all humans, whatever their race or ethnic group, share the same DNA (actually we share about 98% with chimpanzees – our closest animal relative). The very small difference of 0.01% in DNA is enough to make each of us individual. From a biological classification point of view, we are all pretty much identical – otherwise we would be classed as a different species. However, recent research suggests that over the past 5,000 years new genetic variants have been appearing at a rate 100 times faster than before in human evolution.

100

A sneeze exhales air and other matter out of the nose at about 100 miles an hour (161 kph) – containing about 40,000 bacteria-laden droplets that can travel as far as 12 feet. In western countries the traditional response to a sneeze is "Bless you", but in many other countries it is akin to "To your health". All derive from the fact that the first symptom of the Great Plague was sneezing; thus, the nursery rhyme "Atishoo, atishoo, we all fall down".

112.8

The loudest recorded snore is 112.8 decibels, which is very nearly the same as a jet engine. One man in eight snores when sleeping, and one in ten grinds his teeth.

120

The heat created by a human body in an hour is the equivalent of 120 watts, similar to a large light bulb or burning 12.5 ounces of coal.

180

Miles an hour (290 kph) – the speed nerve impulses can travel in the human body. However, this depends on the response needed. For example, if you stub your toe you feel the pressure immediately but the pain a little later, because those particular pain signals travel a fraction slower.

450

Lymph nodes exist in a normal adult body – of which 60-70 are in the neck, about 100 in the thorax, and about 250 in abdomen and pelvis. A lymph node is an oval/kidney-shaped gland, 0.04-1 inch (0.1-2.5 cm) long.

550

Number of hairs in an eyebrow. Their main function is to prevent moisture, mostly sweat, from dripping into the eye. The angled direction of the hairs helps the moisture to flow sideways, away from the eye.

1,200

A full head of human hair will have as many as 1,200 hairs of all sizes per square inch on a scalp (bearing in mind that many are very fine or almost invisible) – almost as many as a chimpanzee. A woman's hair mass is about half that of a male. A man's facial hair grows faster than on any other part of the body, and, if unshaven, would grow to about 30 feet during a lifetime. Every day, a person loses up to 60-100 strands of hair.

1,500

Would you rather be trampled on by a 6,000 lb (2,700 kg) elephant or by a 100 lb (45 kg) woman walking in high heels? Which one exerts more pressure per square inch (psi)? The surprising answer is that elephants, unlike humans, walk with two feet on the ground at a time, and as each foot is about 40 square inches, it delivers a pressure of 75 lbs per square inch (psi). In contrast, a stiletto heel has an area of about 1/16th of a square inch, and thus the woman "wins" easily, exerting a heel pressure of 1,500 psi – and ruining many floors in the process.

2,000

The human body has enough phosphorus to make 2,000 match heads. It also has sufficient carbon to make 900 pencils, sulphur to kill all the fleas on an average dog, and enough fat to make 7 bars of soap.

3,699

Daily calories consumed by the average American. A man should ideally consume 2,500 calories and a woman 2,000.

6,000

Estimates vary considerably as to how many words we speak in a day. Some research attributes 6,000 to 8,000 words to women and only 2,000 to 4,000 to men, while others put these numbers much higher. On average, we have 30 conversations a day.

8,000

The sensitive nerve ends in a clitoris. By contrast, a penis contains only 6,000 – a somewhat unfair distribution if measured on a square inch basis.

10,000

The human body creates anywhere between 1 to 3 pints (0.6-1.7 l) of saliva every twenty-four hours, which amounts to an amazing 10,000 gallons in an average lifetime – just marginally more than the capacity of a petrol tanker.

19,000

Some perfume experts - 'noses' - can detect as many as 19,000 different odours.

33,000

Number of known human illnesses.

50,000

The cells in your body that you will lose, only to be replaced with new ones in the time you have read this sentence.

60,000

Miles (96,600 km) of arteries, veins and capillaries contained in the human body.

100,000

Brain cells a person loses each day after the age of 30 – luckily, we have 10 trillion to spare.

1,900,000

Miles (3,057,100 km) – the distance of all the nerve connections in the brain if they were stretched out.

5,000,000

Hairs on the body, of which only 100,000 or so are on the head.

10,000,000

Pieces of information that the human brain is processing at any one moment through the eyes, ears, touch receptors in the skin, nose and tongue.

75,000,000

The average amount of semen in a human male's ejaculation is between 0.05 and 0.2 fluid ounces (1.5-6 ml). This can contain between 75 million and 900 million sperm, which may be ejaculated at up to 28 miles per hour. The actual orgasm can last from a few seconds to a minute, and the ejaculation consists of 10 to 15 contractions – each lasting about 0.8 seconds.

2,500,000,000

The human heart beats, on average, 2.5 billion times in a lifetime – based on 70 beats per minute for 70 years.

100,000,000,000

The human brain consists of 100 billion neurons, and the brain receives 20% of all oxygen we breathe. In a single day the brain generates more electrical impulses than all the world's telephones collectively.

0

For many years the Australian penal colony was known as 'Botany Bay'. In fact, no prisoner was ever landed in the bay, which was first discovered in 1770 by Captain Cook and the botanist Joseph Banks. Although the destination of the first convicts eighteen years later was to be Botany Bay, as soon as the 'First Fleet' arrived it was realised that it was not a suitable penal colony, and the ships moved on a few miles to Sydney Cove.

0

In the 1962 film, *The Birdman of Alcatraz*, Burt Lancaster plays the part of real life prisoner Robert Stroud, who supposedly gains solace from looking after birds in his cell at Alcatraz. In reality, Stroud never kept any birds at Alcatraz. However, he did have over 100 at Leavenworth Prison, although this caused considerable problems for the prison staff because of the thousands of letters he received as a result of the authorative books he wrote on ornithology.

1/6th

The value that Russians pay each year in bribes is estimated to be $319 billion, or about 1/6th of the country's gross domestic product.

.9

The homicide rate in Ireland is 0.9 per 100,000 of the population. In comparison, it is 1.4 in the UK, 5.5 in the USA, 20.15 in Russia, 32.41 in Jamaica, 49.6 in South Africa and 61.28 in Colombia.

1.57

Albert Pierrepoint (1905-1992) was the most famous, and prolific, of the three members of his family who were Britain's Chief Executioners in the 20th century. (Albert, along with his uncle Thomas, was also the executioner for the Republic of Ireland). When he was appointed an Assistant Executioner in 1932, he was paid the equivalent of £1.57 per execution, with the same amount paid two weeks later – 'provided his conduct was satisfactory'.

When he retired in 1956, over 450 executions later, he was then paid £15 for each.

3

Going under a number of names, the 'Three-card trick', 'Follow the lady', or the 'Three-card Monte' is a confidence game in which victims are bet that they find a specific card – one of three placed face down. Through a number of ploys, manual dexterity and 'distractions' by other spectators, victims hardly ever win, or, if they do, it's only to sucker them to place even bigger bets that they will eventually lose.

3

'The third degree', the toughest level of interrogation by police or authorities – often implies physical violence or extreme mental cruelty. The origins are uncertain, but one explanation is that the 'first' degree was the arrest, the 'second' was taking the suspect to the place of confinement and the 'third' the actual questioning.

3

Because everything had to arrive by boat, it cost three times as much to run 'the escape-proof' island prison of Alcatraz, in the middle of San Francisco Bay, compared with any other US prison – resulting in its closure in 1963.

3

The Chinese criminal organisation The Triads originally started in the 1760s as a patriotic resistance to the Manchu Emperor. The name means 'The Three Harmonies of Society' – referring to the unity between Heaven, Earth and Man. Although outlawed for many years, it was only after the collapse of the Chinese Imperial dynasty in 1911 that they resorted to organised crime as we know it today.

4

In England and Wales, the High Court and the Court of Appeal have four separate 'Sittings' or 'Terms' in a year. Their dates, originating from medieval times, were geared so there were no sittings during Lent, major Christian festivals or harvest. Nowadays the terms run as follows – Hilary: from early January to the Wednesday before Easter; Easter: from the second Tuesday after Easter Sunday to the

Friday before spring bank holiday; Trinity: from the second Tuesday after the spring bank holiday to 31 July; and Michaelmas: from 1 October to 21 December.

5

There were five confirmed murders by 'Jack the Ripper' between 31 August and 9 November 1888 (the victims' graves still exist). There were another 13 alleged, but not fully attributed victims between 26 December 1887 and 29 April 1891.

6

Of the Bible's Ten Commandments, the sixth is 'Thou shalt not kill'. The eighth Commandment states 'Thou shalt not steal', and the ninth 'Thou shalt not bear false witness'.

8

The IRA's Patrick Magee received a punishment of 8 life sentences after he was found guilty in the British Courts of bombing the Grand Hotel in Brighton in 1984, trying to kill Prime Minister Margaret Thatcher and her Cabinet. The bomb killed 2 men and 3 women. Magee was released 15 years later under the terms of the Good Friday Agreement.

8

People killed by handguns in Britain in 1970. By 2006/07 the number of fatalities had risen to 55. Overall, there were 10,182 gun-related offences. However, this is under .5% of all crime.

8

The average measurement of an American prison cell is 8 feet wide, 12 feet long and 8 feet high.

10

10, Rillington Place, Notting Hill, London, was the site of a series of notorious murders by John Christie between 1949 and 1953. They were subsequently made famous through Ludovic Kennedy's 1965 book, and then by a film in 1970. To escape its notoriety, Rillington Place was renamed Rushton Close, and then knocked down and redeveloped in 1981 under a new name – Bartle Road.

10

In 1933, ten minutes separated the time of birth of the infamous East London criminals, the Kray twins – Reggie first, and then Ronnie. They came from a mixture of Irish, Jewish and Romany stock, and at school a teacher commented about them "The twins were never the slightest trouble to anyone who knew how to handle them". This changed as they became leaders of London's organised crime – involved in armed robberies, protection rackets and arson, as well as the murder of Jack 'The Hat' McVitie. Eventually arrested by Detective Superintendent 'Nipper' Read, they were both sentenced to life imprisonment in 1968. Ronnie died in 1995 and Reggie in 2000. Their mother Vi always insisted "They were such good boys".

11

The critical measurements of the human body used by Frenchman Alphonse Bertillon to identify criminals. After adoption in many countries, this 'anthropometric system' was, because of a number of errors, superseded by far more reliable fingerprinting.

13

The 'Yorkshire Ripper', Peter Sutcliffe, is known to have killed 13 women between 1975 and 1980. Despite an extensive police investigation to find the killer of these women, Sutcliffe was arrested in 1981 only on the grounds of his car having false number plates. On subsequent questioning, it was realised that Sutcliffe fitted the police profile of the murderer, and he then admitted to the murders.

14

Children under the age of 14 are not allowed admission to watch the proceedings in the Old Bailey, the UK's Central Criminal Court. The Lord Mayor of London and Aldermen of the City of London are traditionally entitled to sit on the judge's bench during a hearing, but they are not permitted to take part.

15

There is an unusual symmetry in the fact that there are fifteen letters in the names of each of the assassins of two famous US Presidents – Abraham Lincoln (by John Wilkes Booth) and John F. Kennedy (by Lee Harvey Oswald). Other coincidences; Lincoln was elected in 1860, Kennedy in 1960; Booth was born 1839, Oswald 1939. And both Presidents were succeeded by Johnsons – Andrew, born 1808, and Lyndon 1908.

35

Most police chases are conducted at high speed. Not so in Los Angeles on 17 June 1994, when suspected murderer O.J. Simpson's car drove at a sedate 35 mph (56 kph) for about 50 miles followed by a procession of a dozen police cars, reporters and helicopters. Such was the media interest in the pursuit that NBC broke off transmission of the National Basketball Finals to cover the story.

37

The death penalty is authorised in thirty-seven US states, as well as the Federal Government and the US military. Even though Nebraska is one of these states, its Supreme Court ruled in 2008 that the 'electric chair' violates its constitution, and thus, without any alternative, it is left without a legal means of carrying out a death sentence.

38

Members of the Scottish clan MacDonald were slow in pledging allegiance to the new English King, William of Orange, so on 13 February 1692, thirty-eight of them were killed by the supposedly friendly neighbouring Campbell clan, whom they had welcomed into their homes. A further 40 or so MacDonald women and children died from exposure when their houses were subsequently burned. Universally known as the 'Massacre of Glencoe', this event still stands as a symbol of abused hospitality.

50

One of the simplest scams ever – with a 50% success rate. An Australian conman advertised that if you sent him $100, he would 'predict the sex of your next baby – and if he got it wrong, then he'd 'guarantee you'd get your money back'. He was on to a winner. He always predicted a 'boy', as, in reality, it didn't matter because he was likely to be right 50% of the time, so he could afford to reimburse those for whom he had got it wrong.

52

The Russian serial killer, Andrei Chikatilo – the 'Butcher of Rostov', the 'Red Ripper' or the 'Rostov Ripper' – was convicted of murdering 52 women and children between 1978 and 1990. The only way he could achieve sexual satisfaction was when he was in the process of stabbing his victims. A moribund police investigation into the murders meant Chikatilo roamed free for some years until he eventually fell under suspicion, was interrogated and subsequently confessed. He was executed by a pistol shot to the head in 1995.

60

60, Andrassy Street, Budapest, was the much-feared secret police centre, whose murderous excesses contributed towards the 1956 Hungarian uprising. Nowadays, the street is a beautiful and popular shopping avenue, but number 60 has become a museum with the roof overhang spelling out the word 'TERROR'.

193

Just how many more murders can there be in the tranquil rural setting of the TV series *Midsomer Murders*? Up to the beginning of the 11th series in mid-2008, Chief Inspector Barnaby, played by John Nettles, has investigated an amazing 193 local murders in the 60 episodes since 1997. A further twelfth and thirteenth series are planned – so viewers in the 204 countries where it is broadcast can expect the body-count to keep rising.

250

It is thought that the probable total of victims killed by Dr Harold Shipman from Manchester was around 250 (about 80% were women). He was, perhaps, the most prolific personal serial killer of all time, and was officially convicted for just 15 sample murders, received 15 life sentences – but committed suicide in jail in January 2004.

547

In the USA, the self-styled 'Son of Sam' serial murderer, David Berkowitz, was found guilty for murder in 1978, and is still serving the total sentence of 547 years in jail for the killing of at least 6 people.

914

In 1978, 914 men, women and children of the mysterious 'People's Temple' cult were found dead in Guyana after a mass suicide and murder instigated by its leader, Jim Jones, who had convinced his followers to drink a cyanide-laced soft drink. He committed suicide by shooting himself. Jones was fundamentally a Marxist, and when the cult was first formed in America in the 1960s, he recruited his followers to practise what he called 'apostolic socialism'. Having been based in several US states, in 1974 he moved his whole cult to an agricultural community in Guyana, but there were increasing public accusations of abuse. The night before the mass suicide, a US Congressman, three journalists and a defector were shot after visiting the camp.

1,000

From about 900 AD, the self-explanatory 'death by a thousand cuts' was a Chinese form of execution for particularly serious crimes. This punishment worked on three levels – there was public humiliation; it was slow and lingering; and, on Confucian principles, the body was not 'whole for its after-life'. This punishment was officially banned in 1905.

1,208

Built in 1880, London's Wormwood Scrubs Prison has 1,208 cells and can accommodate 1,256 prisoners.

1,400

In America's maximum security prison, Supermax, Florence, Colorado, there are at least 1,400 remote controlled steel doors, along with detector pads and clear lines of sight for the guns in the watchtowers. Built in 1994, Supermax cost $60 million. 23 hours a day in solitary confinement in Spartan cells is still a way of life for the inmates of such prisons.

1720

In 1720, the year of the 'South Sea Bubble', speculation was so rife in the City of London that fraudulent companies were easily able to attract investors. One such put out a prospectus that stated his venture was 'a company for carrying on an undertaking of great advantage, but nobody to know what it is'. Eager investors besieged the office and paid £2 deposits. By three o'clock the promoter had closed his doors and left for the Continent with £2,000 (a quarter of a million pounds today), never to be seen again.

1783

From 1196 to 1783, Tyburn – currently Marble Arch – was the primary location in London for the public execution of criminals. Initially, victims were hung from a tree on the banks of the River Tyburn, but later this system was replaced by gallows.

1788

In a debate in 1788, in the French National Assembly, Dr Joseph-Ignace Guillotin suggested an alternative method of executing prisoners. However, strangely, he had no further involvement with the design and development of the eponymous 'guillotine' which was to become the French method of capital punishment. The last public guillotining was in 1939, and the final in prison was that of torturer and murderer Hamida Dijandoubi in 1997.

1964

Although capital punishment remained on the UK's statute books for certain offences until 1985, the last hangings were at 9 a.m. on 13 August 1964 – of Peter Allen at Walton Prison, Liverpool, and Gwynne Evans at Strangeways Prison, Manchester, – both for the murder of Alan West in April that year. The last woman to be executed was Ruth Ellis on 13 July 1955.

1964

As in the UK, the death penalty in the Republic of Ireland was an available sentence until 1964 for the crimes of treason, murder, piracy with violence and war crimes. After then, capital punishment was only available in cases of treason and some murders, such as that of a Garda (police officer) and heads of state. Capital punishment was finally abolished by Ireland in 1990, and in April 1954 the last man to be hanged was Michael Manning – for murder.

2122

The number in North Clark Street, Chicago, where the 'St. Valentine's Day Massacre' of seven rival gangsters occurred on the orders of Al Capone in 1929.

5,000

...And in the same year, five thousand bullets were fired at Al Capone's Chicago headquarters by 50 gangland rivals in the 'Bootleg Battle of the Marne'. Capone was unhurt.

10,000

...And four years earlier, in 1924, Al Capone sent the biggest wreath to the funeral of Dion O'Bannion, the Chicago gangster, whose murder he had ordered, and which was attended by 10,000 people. As O'Bannion was a major crime figure, the Catholic Church initially denied him burial on consecrated ground. This killing started the gang warfare that ultimately led to the 'St Valentine's Day Massacre'.

22,436

The total number of Polish prisoners estimated to have been killed on Stalin's orders in 1940. Amongst the 15,131 executed at the infamous Katyn Wood and two other locations were almost half the Polish officer corps, including Admirals, Generals, and Colonels, as well as university professors and physicians, together with writers, journalists and other intellectuals. Stalin was cynically removing Poland's potential leaders.

41,000

US dollars – the extortionate price of one broom charged to the City of New York at the height of 'Boss' Tweed's (1823-1878) amazingly corrupt control of New York City's finances in 1870 through his Tammany Hall political machine (as featured in the film *Gangs of New York*). It is estimated that Tweed plundered the City's coffers by between $75-200 million dollars.

80,000

Between 1852 and 1946, about 80,000 prisoners were shipped to the French disease-infected penal colony in French Guiana – part of which was the infamous Devil's Island. Probably the most famous prisoner was Captain Alfred Dreyfus, who had been wrongly convicted of treason in 1895.

88888

When rogue trader Nick Leeson took unauthorised speculative positions on behalf of Britain's oldest merchant bank, Barings, he hid his transactions in an unused error account, numbered 88888. Right from the start in 1992 the account showed a loss, and the bank's management remained blithely unaware of this until 1995 when Leeson fled, leaving a terminal £875 million hole in the bank's balance sheet. In 1993, Leeson had been the Baring's 'blue-eyed boy' as his official trading on behalf of the bank made a surplus of £10 million – about 10% of the bank's profit for that year.

100,000

People killed by 'Vlad the Impaler', (1431-1476), the undeniably cruel ruler of Walachia and defender of his kingdom against the Ottoman Turks. Although impaling was his favourite method of execution, he devised many other cruel methods of dispatching those who displeased him. Vlad Dracul (The Dragon) provided the inspiration for author Bram Stoker's (1847-1912) horror story, *Dracula* (1897).

125,000

Victor Lustig was a remarkable conman. His best scam was in 1925, when under the alias of being the French 'Deputy Director–General of Posts and Telegraphs', he managed to sell the Eiffel Tower to a scrap metal merchant for $125,000 – at the same time receiving a profitable backhander. Because his very embarrassed victim kept quiet, Lustig had the almost unbelievable nerve to try to repeat the scam again that year, before fleeing to America.

200,000

The legendary D.B. Cooper perpetuated one of the cleverest hijackings ever. He threatened to blow up a Boeing 727 in 1971, extorted $200,000 from the airline Northwest Orient, and after the plane took off he parachuted out with twenty-one pounds in weight of $20 bills strapped to his body. The case has subsequently baffled law enforcement agencies, because D.B. Cooper was an alias – so nobody knows the hijacker's real identity, nor is there any record of his landing – so it is still uncertain as to whether he survived or not. What is more, none of the money has been recovered.

247,000

The area in acres (100,000 hectares) in Colombia which is under coca cultivation. Despite energetic Government efforts to eradicate the crop, this area grew by 27% in 2007. Taking into account the two other main coca-producing countries, Peru and Bolivia, overall production has grown by 16%.

829,627

Last year in the USA 9 out of 10 of 829,627 marijuana arrests were for possession - not for sales or trafficking. That's one pot-related arrest every 38 seconds. An all-time record.

1,600,000

The sum of 1.6 million Dutch guilders was the price achieved by Han van Meeregen when he sold a Vermeer painting to Nazi Reichsmarschall Goering during WWII. After the war, Goering's art hoard was confiscated, and art experts were bemused by the wonderful Vermeer, of which they had no previous knowledge. The sale was traced back to van Meeregen, who was now in serious trouble, as his transaction was considered a treasonable act of collaboration. Much to their surprise, he was able to prove that it was a forgery – perfectly painted by himself – as were dozens of other 'masterpieces'.

2,258,483

Prisoners in US jails at the start of 2007; this is 501 per 100,000 of the population, and five times the population of Luxembourg! Of these, 112,498 are women. In Russia, the prison population is just under 500 per 100,000 of population.

12,000,000

New York's property tycoon and hotelier 'The Queen of Mean', Leona Helmsley, was jailed for tax evasion in 1995 after declaring "Only the little people pay taxes". When she died, her dog 'Trouble', a fluffy Maltese bitch, was left a substantial $12 million legacy so it could live a life of luxury. Even though Leona Helmsley's estate was an estimated $4 billion, 'Trouble' received more than any of her human relations.

20,000,000

Although the figures vary considerably, at least 20 million

Russians were imprisoned in the Gulag, which comprised 476 camp complexes and perhaps thousands of forced labour camps, during Stalin's Great Terror from 1930 until his death in 1953. Gulag was actually the Russian language acronym for The Chief Administration of Corrective Labour Camps and Colonies – whose department for penal settlements was part of the State Security, and started in 1919.

50,000,000

At a cost of £50 million, 'Operation Crevice', which brought to court suspected Islamist radicals charged with conspiracy to detonate bombs in the UK, was one of the most expensive criminal cases in the history of British criminal justice. It was also one of the longest, lasting over one year. It required 50 homes, cars and business premises to be searched, 80 computers to be seized and 18 people to be arrested. There were 105 prosecution witnesses and 173 interviews, with the total involvement of 960 police officers.

3,600,000,000

The £3.6 billion lost through fraud by Jerome Kerveil, the rogue French banking trader working for Société Générale in 2008, which makes the £875 million lost by Barings Bank's Nick Leeson in 1995 seem relatively modest.

6,000,000,000

The FBI estimates the worldwide value of art theft to be in the region of $6 billion. It is difficult to put an exact price on the most valuable paintings ever stolen, but, at £80 million, the four Impressionist paintings by Cézanne, Degas, Van Gogh and Monet from the Emile Bührle Foundation in Zurich, Switzerland, have to be amongst the highest. (The Monet and the Van Gogh were recovered on the back seat of an unlocked car in a mental hospital a few hundred yards away). Other major art thefts include $60 million worth of Picasso's work stolen from his granddaughter in Paris in 2007, and Edvard Munch's *The Scream* from a Norwegian museum in 2004 (and recovered in 2006).

0

By the end of World War II, the Germans had no spies still working for them in Britain, as they had all been caught or been 'turned' to work for the Allies.

0

The Blair/Brown Cabinets from 1997 are the first in which none of the members had ever served in the Armed Services.

0

Despite its designation, no horses were used by the 1st United States Volunteer Cavalry when it charged into battle in Cuba during the Spanish American War in 1898; it served as dismounted infantry, and was more famously known as Theodore Roosevelt's 'Rough Riders'.

0

In 1933, for the first time ever amongst British Army regiments, the Gordon Highlanders achieved a zero rate of illiteracy amongst all ranks.

0

The Mitsubishi 'Zero', the most famous Japanese fighter plane, which ran rings around Allied opposition in the early stages of WWII.

0

The traditional reason given for the fall of Singapore to the Japanese in 1942 was that the main defensive heavy guns had been installed so that they could only face one way in preparation to repel a seaborne attack. In fact, no guns 'pointed the wrong way'; all could be directed towards the land side as well, but as they were flat-trajectory anti-ship guns, they were useless against the advancing Japanese infantry in the surrounding jungle.

.303

Inches. The calibre of the British and American infantry rifles in World Wars I and II.

.45

Colt pistol – the most famous handgun in the American West – the six shooter pistol designed by the Colt Manufacturing Company for the US cavalry, and adopted in 1873.

1

From the 16,500 soldiers and camp followers that set out to retreat from Kabul in 6 January 1842, only one survived – Dr William Brydon, who reached Jalalabad 90 miles away on 13 January. Part of his skull had been sheared off by an Afghan sword, and Brydon survived only because he had stuffed a copy of *Blackwood's Magazine* into his hat to fight the intense cold weather. The magazine took most of the blow, saving the doctor's life. In 1857, he survived a second dramatic siege – that of Lucknow, during the Indian Mutiny.

1

The US had only one spare atomic bomb in reserve in 1945 after Hiroshima and Nagasaki, although the Japanese thought there were many more and duly surrendered. In fact, atomic bombs have been only ever used on these two occasions – thankfully making them one of the military products least used for their purpose.

3

Days, the miraculously short time it took to repair the American aircraft carrier USS *Yorktown*, after the Battle of the Coral Sea in May 1942. If it had taken the estimated 90 days, the Battle of Midway would not have been won – nor perhaps the Pacific War.

3

Of the four sons of twenty sixth US President Theodore Roosevelt, three died as a result of war. Quentin, a pilot with the 95th Aero Squadron, was shot down in 1918 in France. He was later re-buried next to his brother, Theodore (Ted jnr.) who had died of a heart attack after leading an assault in Normandy in 1944, and finally Kermit, who, after a distinguished combat record in WWI (including being awarded a British Military Cross), killed himself in Alaska in 1943. The fourth son, Archie, survived, having been badly wounded in France in 1918, was awarded the Croix de Guerre, and then fought again with distinction in WWII.

3

The French General, Maxime Weygand, predicted that "England would have her neck wrung like a chicken within three weeks" by the Germans at the beginning of World War II. As Churchill later riposted "some chicken, some neck".

3.75

The final height of Captain Jeffrey Hudson was only 3 feet 9 inches (1.1 m). Initially, he was only 18 inches high (0.45 m). He became the 'Queen's Dwarf' at the court of Charles I, carried out a number of diplomatic missions, was appointed a 'Captain of Horse' and fought as a Royalist in the Civil War. Hudson was subsequently unlucky enough to be captured in 1643 by Barbary pirates, and it was during his twenty-five year captivity that his height doubled.

4

During WWII, Henry J. Kaiser's shipyards took only four days to produce a complete 'Liberty Ship' (a merchant and cargo boat) by highly efficient assembly line methods. The keel for the 10,500 ton *Robert E. Peary* was laid down on 8 November 1942, and was launched 4 days and 15.5 hours later.

4

It took Napoleon only four hours to send messages from Paris to Rome by his semaphore system.

IVC

The Oflag number of Colditz Castle – the famous WWII prisoner of war camp, supposedly escape-proof. However, over the five years of the war, there were over 300 escape attempts, in which 120 actually got away but were recaptured, while thirty-one made it back from the castle, the hospital and the grounds to their home country.

5

Israeli fatalities in the daring 2,500 mile raid on Entebbe, Uganda, in 1976, to rescue hijacked Jewish hostages – three hostages killed in crossfire, an ill Mrs Block, later murdered on the orders of President Idi Amin, and Lt. Col. Yoni Netanyahu, the leader of the rescuing force.

5

The future US General George Patton came fifth in the 1912 Olympic Pentathlon. Having come third in athletics, sixth in equestrianism, and fourth in fencing, he would have won a medal had he not insisted on using a large 'manly' service pistol in the shooting where he came only twentieth, because the judges had difficulty marking the very large size of bullet holes on his target.

6

In 1863, six French Foreign Legionnaires charged 2,000 Mexican troops at the end of the Battle of Camerone. Three of them survived. 'Camerone Day' is remembered every year on 30 April, and central to the celebration is the preserved wooden hand of the commander of the heroes – Captain Jean Danjou.

6

Six US Presidents have been professional soldiers; Washington, Jackson, Harrison, Taylor, Grant and Eisenhower.

7

It was seven hours and forty minutes between the first sighting of the French Fleet and the firing of the first shot at the Battle of Trafalgar, (1805). The French fleet had thirty-three ships while Nelson's had only twenty-seven.

7

Michael Collins, Ireland's legendary Nationalist leader, was seven minutes late for the parade which at long last handed over the government to Ireland in 1921. When the British Lord Lieutenant complained, Collins riposted "You had to wait seven minutes, but we had to wait seven hundred years ".

8

Wagons of tools and charcoal that engineer General Eblé kept against Napoleon's orders, and which subsequently enabled him to build two bridges across the Berenzina River, thus saving the remnants of the Grande Armée on its disastrous retreat from Russia in 1812.

8

Years – the length of the longest modern war, between Iraq and Iran (1980-1988), much of which was fought in a similar way to Europe's WWI – trench warfare, bayonet charges, barbed wire, and the use of gas. It resulted in a combined total of almost one million dead.

9

'Cat-o'-nine-tails', a nine-roped flogging whip. The condemned sailor had to make it himself, and when it was produced the next day, it was called 'letting the cat out of the bag'.

11

The highest amount of VCs awarded in any one action – the defence of Rorke's Drift in Natal, South Africa, in January, 1879. Of the garrison of 139, only 104 were fit to defend it against the 4,500 strong Zulu 'impi'. At the end of the battle, the Zulus had lost nearly 500 and the defenders only 17 – quite a few of whom were shot by captured rifles, rather than killed by the traditional Zulu 'iklwa' short stabbing spear.

13

The British intelligence had 'MI' (Military Intelligence) departments numbered 1 to 19, with the exceptions of MI13 and MI18 which never existed. Each had a specific function; for example – MI1 which was responsible for code breaking, MI4 for aerial reconnaissance, MI16 for scientific intelligence and MI19 for obtaining intelligence from enemy prisoners of war. Nowadays, only two remain – MI5, the counter-intelligence and security service, and MI6, the external intelligence agency.

15

The number of times Gibraltar has been besieged since it was ceded in perpetuity by Spain to Britain in 1713 by the Treaty of Utrecht.

16

About 3,500 men and women were arrested after Ireland's 'Easter Rising' in 1916. Sixteen of their leaders were shot by firing squad by the British, including the wounded James Connolly who was strapped to a chair, and Joseph Plunkett, executed ten minutes after being married to his fiancée. This turned an unpopular revolt into martyrdom, and ultimately ensured Ireland's independence.

16

The extraordinarily young age at which Boy First Class 'Jack' Cornwell won his posthumous Victoria Cross on 31 May 1916 at the great sea Battle of Jutland. His gun turret had been destroyed by an enemy shell, and all its crew killed. However, even though he was mortally wounded, Cornwell stood resolutely alone at his gun position until the end of the action, loyally awaiting further orders.

17

The number of the Arctic convoy that sailed to Russia in 1942, PQ17, which suffered the worst loss – 25 merchant vessels sunk out of 36.

19

In the 1930s just before World War II, the US Army was nineteenth in size ranking among nations, just smaller than that of Portugal.

20

Before World War I, only twenty books a year in Britain were published on warfare. This is in contrast to the seven hundred published in Germany.

20

In order to escape immediate burns from a 1 megaton thermonuclear explosion you would need to be at least 20 miles away. 2.5 miles is the distance from a 1 kiloton nuclear blast where you would not be knocked off your feet.

22

The number of kills that air ace Hermann Goering achieved leading Richtofen's 'Flying Circus' in 1918, winning Germany's highest military decoration, the 'Pour le Mérite' (also known as the 'Blue Max') – its French name dates back to 1740 when it was introduced by Frederick the Great, the common language of European royal courts being French. Later, as a leading Nazi and head of Hitler's Luftwaffe, Reichsmarschall Goering was rather less heroic.

30

Rather than committing their main armies during the struggle for the Duchy of Brittany, unusually only thirty French and English knights and squires were chivalrously chosen to fight on each side in 'The Battle of the Thirty' in 1351. The French were the victors.

34

In 1941, Russia's T34 tank gave the invading Germans a terrible shock. Simple, rugged and fast, it served until the mid 1950's, giving another shock to the Americans in Korea.

34

The number of battles fought over time at strategic Megiddo in Israel – perhaps better known as Armageddon.

37

Eton College, Britain's famous public school, has produced thirty seven winners of the highest British award for bravery, the Victoria Cross. The first four Old Etonian recipients were all awarded their medals during the Crimean War. The most recent was the posthumous award to Colonel 'H' Jones during the Falklands War in 1982. The first Old Harrovian to receive the VC was Sir Robert Peel's son, Captain Sir William Peel, who died during the Indian Mutiny.

39

Britain's allies in the World War 1 – Albania, Andorra, Armenia, Australia, Belgium, Britain (including Ireland), British Crown Colonies, Bolivia, Brazil, Canada, China, Costa Rica, Czechoslovak Legions, Cuba, Ecuador, France, Greece, Guatemala, Haiti, Honduras, the Indian Empire (now includes Bangladesh, India and Pakistan), Italy, Japan, Liberia, Montenegro, New Zealand, Newfoundland, Nicaragua, Panama, Peru, Portugal, Romania, Russia, San Marino, Serbia, Siam (Thailand), South Africa, the United States of America and Uruguay. The Central Powers comprised Austria-Hungary, Bulgaria, the German Empire and Turkey's Ottoman Empire.

40

During the German 'Blitz' on London in 1940, bombs weren't the only danger. The use of lights was forbidden and this total blackout meant that, each night, forty pedestrians were knocked down by traffic and other hazards.

47

1947 was the year that Mikhail Kalashnikov designed his legendary AK-47 assault rifle, the most ubiquitous weapon in the world with over 100,000,000 having been produced. The weapon's simplicity and reliability explain its success.

50

Fifty per cent of American casualties in the Far East in WWII were due to malaria.

60

Animals have their own bravery award, the 'Dickin' medal – instigated in 1943 and named after Mrs Maria Dickin, (1870-1951), founder of the The People's Dispensary for Sick Animals. It has been awarded 60 times to acknowledge the work and bravery of animals in war – 34 times to carrier pigeons. It is traditionally presented by the Lord Mayor of London.

77

Miles from Paris, a huge German gun fired a shell every 20 minutes into the city for 139 days in 1918. It killed 1,000 civilians and increased hatred towards Germany. An expensive masterpiece, with each shell individually numbered, needing to be a little longer and fatter to allow for the wear of the barrel, which, after only 65 shells, had to be replaced.

80

The German First World War fighter ace Baron Manfred von Richtofen, 'The Red Baron', shot down 80 enemy aircraft. However, on 21 April 1918, while on a sortie over British lines, he was shot through the heart by either Australian riflemen on the ground, or by Canadian air ace Arthur Brown. Richtofen's body was retrieved from his crashed plane by British forces, and he was buried with full military honours.

80

The number of 'good troopers' that Captain William Fetterman boasted he would need 'to ride right through the Sioux nation'.

In 1886, Crazy Horse lured him into a trap, and he and exactly 80 troopers died.

82

During each year of the long drawn-out Napoleonic War, disease or accidents on board British Navy ships caused about 82% of the deaths, rather than enemy action.

88

During the Spanish Civil War the Germans first discovered how effective their legendary 88 mm anti-aircraft gun was going to be in WWII, and that it was equally potent against tanks. It was later installed on their much-feared Tiger tank.

90

In 1862, the inventor John Ericsson took only 90 days during the American Civil War to deliver the Union forces' first ironclad ship, the USS *Monitor*. She was only just in time, a matter of hours, to fight the Confederate Ironclad, *Virginia* (*Merrimac*), in a pivotal battle to stop her destroying the blockading Union wooden ships.

100

The British used 100 octane fuel in their aircraft in WWII, giving them an edge over the 87 octane used by the Germans.

100

The famous '100 Days' (actually 116) in 1815 sent Europe into temporary turmoil. It was the period between Napoleon escaping captivity on Elba, gathering his loyal troops around him and then finally losing the Battle of Waterloo on June 18th.

100

The number of troops a Roman Centurion commanded, roughly the equivalent command of a modern infantry Major.

100

The small number of power stations that, if hit hard by Allied bombing, would have closed down Germany in 1943, thus shortening the war. If just two had been destroyed, then this would have brought Berlin to a standstill. For some reason, they were never targeted.

103
The number of Royal Air Force fighters shot down in one day above the Dieppe Raid in 1942, the RAF's single greatest defeat.

109
The Messerschmitt BF109, the longest-serving German fighter of WWII, of which 30,000 were built from 1937-1945. Veteran Spanish Air Force versions flew in the 1969 film *The Battle of Britain*.

109
The number of the torpedo boat commanded in the Pacific Ocean in WWII by young American President-to-be, John F. Kennedy, was PT109. In 1943, while patrolling in the Russell Islands, his boat was cut in half by the Japanese destroyer, *Amagiri*, travelling at 40 knots. Kennedy helped the surviving crew to swim to the nearest island, and then on to further islands until they found help. He was later awarded the Navy and Marine Corps medal for bravery, and the Purple Heart for injuries sustained that night.

110
Hours, the limit of flying training time in WWII for German new pilots, reduced by lack of fuel in 1944. That same year, British pilots were given 340 hours training and Americans 360.

116
The actual length of the so-called 'Hundred Years War', the longest war in history, from 1337-1453. It featured the three 'long-bow' English victories, Crécy, Poitiers and Agincourt, and the drama of Joan of Arc.

130
Spanish ships in the Armada of 1588. 67 ships from the fleet were lost, 28 of which were wrecked on the Irish coast, largely because the Spanish did not realise that the Gulf Stream current was driving them eastwards.

157
The number of casualties per thousand of British troops fighting in Burma in 1943 caused by venereal diseases, compared with 8 per thousand by battle wounds. The same year, casualties among

American white soldiers in Tunisia reached 33.6, and among black soldiers an amazing 451.3. The US forces eventually issued condoms, but prudishly insisted to the public back home that they were really for 'protecting machine-gun barrels'.

183

In March 1836, 183 Texans under the command of Jim Bowie (of eponymous 'knife' fame) and including famous hunter and Indian fighter, David 'Davy' Crockett, defended the Alamo Mission in San Antonio from 5,400 Mexican troops in their struggle to make Texas independent from Mexico. They lost the 13-day siege and all died, but their sacrifice was not in vain. Such was the anger and fervour generated by the incident that the Alamo's victor, General de Santa Anna, was beaten and captured some weeks later at the battle of San Jacinto.

200

Out of 693 ships and boats engaged in the rescue of British troops from the beaches of Dunkirk, 200 were sunk. However, the operation allowed 338,226 British and French troops to escape to Britain between May 26 and June 4 1940.

200

The remarkably short, but very active career of battleship HMS *Prince of Wales*, lasted only 200 days. She was commissioned in March 1941, fought the German battleship *Bismarck*, took Churchill to the Atlantic Conference with President Roosevelt and was then sunk by Japanese aircraft in December.

230

The civilised world has been at peace for only 230 years during the last 3,500.

235

George Custer led 235 soldiers to their death at the Battle of the Little Bighorn in 1876, somewhat mistakenly shouting, "We've got them this time, boys!" when he faced 4,000 Sioux and Cheyennes.

252

In 1898, there was a mysterious explosion on the USS *Maine*, which killed 252 of her crew in Havana harbour, Cuba. The flames of the subsequent war with Spain were fanned by newspaper baron William Randolph Hearst's campaign "Remember the *Maine*, to hell with Spain". It also triggered America's expansion into the Philippines – another Spanish colony.

297

The Prussian army lost only 297 from smallpox during the Franco-Prussian War (1870-71). But in the French army, where vaccination was not compulsory, 23,400 died of the disease.

445

In 1852, 445 troops stood to attention in ranks awaiting their fate on the deck of HM Troopship *Birkenhead* as it broke up off the coast of South Africa, while the women and children were put into the very few serviceable lifeboats. Only a few of the 445 survived. This disaster was responsible for the introduction of the maritime protocol 'Women and children first'.

460

In World War I, the Western Front line was 20 miles deep and ran for 460 miles from Belgium to neutral Switzerland. The conflict on this front alone resulted in a combination of 6 million deaths and 12 million wounded.

617

This was the designated number of the famous RAF 'Dambusters Squadron' whose 'bouncing bombs' during WWII breached the Eder, Möhne and Sorpe dams, releasing over 300 million tons of water vital to German industry on the industrial Ruhr. The squadron's motto, appropriately and ironically, became 'Après moi, le déluge' ('After me, the flood').

673

The number of men who charged with the Light Brigade on 25 October 1854 was 673. At the end of this epic event 118 were killed, 127 wounded and only 195 still on their horses.Tennyson's immortal lines – 'Not tho' the soldier knew / Someone had blunder'd; / ...Into the valley of death / Rode the six hundred' – would not have rhymed so well if he had used the correct number.

719

The number of British Forces who died in Northern Ireland between 1969 and 2007. 1972 was the worst year, with 146 security members being killed. In comparison 1,109 were killed in the Korean war (June 1950 – July 1953) and 255 in the three months of the Falklands war in 1982. In the first Gulf War (1990-91) there were 49 deaths, Iraq (2003-08) 176, Balkans (1991-99) 65, Cyprus (1955-59) 105, Borneo (1962-66) 131, and Suez (1956) 14.

720

Feet is the standard length of a Royal Naval nautical rope which allowed a ship to anchor in 40 fathoms (240 feet). To make a rope of this length, the initial individual strands need to be 1,000 feet long before they are twisted. In the year of Trafalgar, (1805), Nelson's flagship, HMS *Victory*, used 29 miles of rope of various sizes, and overall the Royal Navy used 16,500 ton of cordage.

727

Victories by German Major Eric 'Bubi' Hartmann, the most successful air ace in history. 345 kills were against the Soviets, 260 of which were fighters. He was in 825 aerial combats and crashed 14 times due to mechanical failure or enemy debris, but was never shot down. In comparison, Britain's top ace scored 34, and America's 40.

800

In the American War of Independence, (1775-1783), the British had a Navy of 800 ships which was opposed by only 30 American vessels. However, despite these overwhelming odds, the Americans won quite a number of victories, the most notable being that of John Paul Jones, later dubbed 'The father of the American Navy', after the *Bonhomme Richard* sinking HMS *Serapis* off Britain's Yorkshire coast in 1779.

1,000

The record military advance on foot, from August 29 to October 8 1941, was 620 miles (1,000 km) – 15.6 miles (25 km) a day – in Russia by the Spanish Blue Division. This was the token force that Franco had lent to Hitler to avoid joining the Axis properly, in a war he realised they were almost certainly going to lose.

1066

A few weeks after the first sighting of Halley's Comet in 1066, regarded as a bad omen by the Anglo-Saxons, William the Conqueror invaded England, and beat them at the battle of Hastings. Hence, the beginning of the Norman domination of Britain.

1,157

The exceptionally high death-rate in WWI of officers – many of whom were educated at public schools – is epitomised by the 1,157 Old Etonians who were killed. The Battle of the Somme alone accounted for 148, of whom more than 30 died on the first day – 1 July 1916. 2,917 Old Harrovians served in the armed forces, of whom 690 were wounded and 644 were killed. This tragic roll of honour would certainly be proportionally replicated in every other school in the UK.

3,400

Each ship of the line at the Battle of Trafalgar (1805) required 3,400 trees to be felled to complete its construction. Made mostly of oak, after the battle quite a number of oak woods were planted so that Britain would never again be short of suitable quantities of wood for shipbuilding.

5,000

Wounded soldiers died unnecessarily in Scutari Hospital during the Crimean War in 1854, because, for all her nursing skills, Florence Nightingale did not understand the dangers of secondary infection.

5,000

At the Battle of Agincourt in 1415, Henry V's total army numbered 6,000, of whom 5,000 were highly skilled longbowmen, and they faced a French army of between 12,000 and 36,000. The English archers had a stock of about 400,000 arrows, which they used to devastating effect. At their fastest rate of fire, each archer could fire 10 arrows a minute – with an effective maximum range of 260 yards (240 m). So, in theory, there could have been as many as 800 arrows a second, or 50,000 a minute, raining down on the French army. If the English had actually fired at this rate, then their supply of arrows would have lasted only eight minutes.

5,000

The greatest loss of tanks in WWII was the 5,000 on ships sunk on the convoys trying to take American and British aid to Russia – together with 7,000 planes which never flew.

7,000

The weight of the concrete Phoenix Units that made up the remarkable artificial 'Mulberry Harbours' was 7,000 tons, and which the Allies towed to Normandy on D-Day in June 1944. Their purpose was to ensure there were immediate facilities to land a steady flow of supplies to support the Allies landing in France, without having to capture any of the heavily-defended French Channel ports.

7,700

When the *Wilhelm Gustloff* was torpedoed in the Baltic Sea by a Soviet submarine in January 1945, it was a maritime disaster five or six times worse than the better known sinkings of the *Titanic* (1912) and the *Lusitania* (1915), as 7,700 of the 10,000 German refugees on board were drowned.

12,000

In WWII, graduates of Russia's Central Women's School for Snipers accounted for twelve thousand Germans killed (top individual score 309). Women made up 800,000 of the Soviet forces, including 40% of combat doctors, and all nurses. More surprisingly, many fighter and bomber pilots and navigators were women, 33 of whom became 'Heroes of the Soviet Union'.

13,500

During the siege of Stalingrad, (1942-43), the Russian secret police, the NKVD, killed well over thirteen thousand of their own Russian soldiers for 'cowardice'. The Red Army issued a total of 800,000 death sentences on its own troops during the first year of the WWII German invasion of Russia.

17,000

Civilians killed by the French army in the Paris Commune in 1871, rather more deaths than they had inflicted on the Prussians who had just defeated them in the Franco-Prussian war.

22,000
By the time of the Armistice in 1919, the British Army on the Western Front had 22,000 carrier pigeons in 150 mobile lofts. The birds were a very effective message-carrying service, as only 2% of those released failed to return.

24,000
From the 22 to the 23 January 1879, the few defenders of the small military outpost in South Africa, Rorke's Drift, fired 24,000 or so bullets to fight off the 4,500 attacking Zulu warriors. They only had 500 bullets remaining when the Zulus retired with 500 dead compared with the defenders' 17.

25,000
German Shepherd dogs, or Alsatians, that died as war dogs in both World Wars. One particular German puppy was saved by an American soldier in France at the end of WWI, and went on to be the Hollywood canine star of *The Adventures of Rin Tin Tin*.

33,000
North Vietnamese and Vietcong deaths in their Tet offensive of 1967. A military failure, it was a propaganda success and America soon left Vietnam.

46,752
US soldiers were killed in Vietnam between 1959 and 1975. However, during the same period, almost twice as many people, 84,633, were killed by guns within America.

51,000
There were slightly more casualties at the American Civil War Battle of Gettysburg, (1863), – 51,000 – than at the Battle of Waterloo, (1815). Wellington's and Blücher's forces combined lost 22,000 killed and wounded, while Napoleon's lost 25,000.

55,000
Many of the Allied troops captured by the Japanese in South East Asia during WII were put to work on building the 258 mile-long Burma-Thailand Railway. This was illegal as it contravened The Geneva Convention. Such were the conditions and their captors' cruelty, that between June 1942 and October 1943, over

55,000 Allied prisoners and Asians died – one for every railway sleeper. By using such a sick and weak workforce, the railway was built very slowly, so it was late and never really of any use to the Japanese war effort.

100,000

Soldiers were addicted to opium by the end of the American Civil War (1860-1865), given to them as a painkiller. With a population of only 40 million, this means that there were proportionally twice as many drug addicts then than today.

200,000

The final death toll of the first use of the atomic bomb at Hiroshima on 6 August 1945.

500,000

One of the most critical obstacles for the British troops to overcome in October 1942, as they started to attack at the Battle of El Alamein, was to clear a route through the German's 500,000 mines laid five miles deep. The minefield was known the 'Devil's Garden'.

500,000

The surprising number of Soviet Army soldiers on horseback during WWII – to good effect. Unlike their German foes, they were not forced to use horses for the lack of fuel, but rather for mobility in a roadless country.

600,000

Americans who were killed in the Civil War. Losses in the English Civil War (1642 – 1651) were probably less than 100,000, including from disease (but excluding Ireland). The Spanish Civil War of the mid 1930s was much more devastating with 500,000 losing their lives, nearly 200,000 of them by execution and murder.

600,000

The deaths in Ireland out of a population of just 2 million, between 1641 and 1652, during Cromwell's subjugation of the country – far worse than the Great Famine of 1845.

765,399

Out of 5.5 million British troops who fought in the First World

War, well over seven hundred thousand were killed.

1,000,000

One million American soldiers were transported safely across the Atlantic to Europe by the *Queen Mary, Queen Elizabeth* and other great liners in WWII, at speeds too fast for U-Boats to intercept.

2,700,000

The number of horses that died as part of the Germans' war effort in Russia in WWII. In comparison, and 130 years earlier, Napoleon's army in Russia lost 200,000 in a quarter of the time.

4,160,000

The British forces fired this number of Vickers shells at the Germans during WWI, which were ironically fitted with German Krupp fuses, under an 1896 licence. Vickers, almost unbelievably, then paid a royalty cheque to Krupp in 1926.

4,860,000

Up to his death in 1227, Genghis Khan (b 1162) and his Mongol hordes had conquered over 4.8 million square miles in China, Russia and Asia.

6,210,000

Poland lost an horrific 6.21 million (600,000 military and 5,600,000 civilian) during WWII – 20% of her population. By contrast, Britain lost 1%, the US .5%, Japan 4%, Germany 11% and the Soviet Union 12%.

6,636,360

Square feet, (29 acres/12 ha) – the total area of the Pentagon building which houses the US Department of Defense, and is occupied by its 24,000 military and civilian staff in 3,705,793 square feet (344,268 sq m) of office space. The total complex covers 583 acres, has a 5 acre courtyard at its centre, 67 acres (27 ha) of parking, 17.5 miles (28 km) of corridors, 284 restrooms, 691 drinking fountains and 4,200 clocks. It has 100,000 miles (161,000 km) of telephone cables and handles 200,000 calls daily, and recieves 1,200,000 pieces of mail monthly.

28,000,000

This staggering number of bottles of wine – along with 2 million bottles of brandy – was taken into Russia by Napoleon when he invaded in 1812. However, his army was 530,000 strong. Factors other than the contents of these bottles actually led to the campaign's disaster!

41,000,000

The approximate total casualties, both military and civilian, in WW1 – 20 million deaths and 21 million wounded.

76,200,000

.5 inch machine bullets were expended by the aircraft of the US 8th Air Force in WWII, which shot down 6,000 German planes, a ratio of 12,700 rounds per kill.

223,000,000

Barrels of oil produced by the United States in 1944. In the same year Germany produced just 6,400,000.

150,000,000,000

To one – the odds estimated by the Germans against an enemy breaking their 'Enigma' code system. However, it was broken at Britain's Bletchley Park, and the decoded messages helped to win the Second World War.

689,000,000,000

The US defence budget has reached $689 billion. Britain's is £33,600,000,000 ($69.3 billion).

0

Times that Rick (Humphrey Bogart) tells the pianist to "Play it again, Sam" in the 1942 film *Casablanca*. Often associated with the film, this is a misquotation. In fact, the relevant dialogue is as follows; when Ilsa (Ingrid Bergman) first enters the Café Américain, she spots old friend, pianist Sam, and asks him to "Play it once, Sam, for old time's sake." When he feigns ignorance, she responds, "Play it, Sam. Play *As Time Goes By*". Later that night, alone with Sam, Rick says, "You played it for her, and you can play it for me" and "If she can stand it, I can! Play it!".

0

World-famous film director Alfred Hitchcock (1899-1980) never ever won an Oscar as Best Director for any of his 66 films for cinema and TV. His fans made a hobby of trying to spot his cameo appearances in his own films – normally very early in the action.

1

Frenchman Marcel Marceau, the world's most celebrated mime artist, contrarily spoke one word in Mel Brook's film *Silent Movie*. His sole verbal contribution was "Non".

1

One-Eyed Jacks – 1961 film starring Marlon Brando, the only film he ever directed.

1

One Flew Over the Cuckoo's Nest – 1975 film starring Jack Nicholson which won 5 Oscars. Its subtitle was 'If he's crazy, what does that make you?'.

1

The One That Got Away – the 1957 film that featured the only German prisoner, Oberleutenant Franz von Werra, to escape from Great Britain in World War II and get back to Germany. The lead was played by Hardy Krüger.

3

Age of Shirley Temple (b 1928) in her first film role. She went on to be probably the most loved of all child stars.

3

Three Coins in a Fountain – 1954 film in which an American girl seeks romance in Italy – based around the Trevi fountain. However, during the film, the audience only sees two coins thrown into the fountain!

3:10

3:10 To Yuma – classic 1957 Western starring Glenn Ford and Van Heflin, remade in 2007 with Russell Crowe.

4

Four Weddings and a Funeral – the British film that consolidated Hugh Grant's pre-eminence as a romantic comedy star, and which unexpectedly grossed $244 million – the highest to date in the British film industry.

4

The Fourth Protocol – 1987 spy thriller film starring Michael Caine, with the future James Bond actor Pierce Brosnan as the Soviet agent.

4

The Four Feathers – 1939 film about apparent cowardice, in which the main character eventually turns out to be a hero. The story has been filmed seven times.

4

The Four Musketeers – the follow-on movie to *The Three Musketeers*, as the producers realised that there was enough footage for two films, and thus created a sequel.

4.20

All the clocks in the film *Pulp Fiction* show the same time.

5

Feet tall (1.5 m) – 1920s Hollywood star Mary Pickford's height. However, Hollywood purposely made her look even smaller and frailer by building oversized sets and furniture. Probably America's first film sweetheart, she was also a very savvy business woman – founding United Artists with her then husband Douglas Fairbanks, and Charlie Chaplin.

5

Five Easy Pieces – 1970 film starring Jack Nicholson, with the tag line 'He rode the fast lane on the road to nowhere'.

5

The Five Pennies – 1959 biopic of jazzman Red Nichols, and his Five Pennies band. Lead star Danny Kaye accurately faked his playing of the cornet, while the real Red Nichols provided the actual soundtrack.

7

Seven Brides for Seven Brothers – this 1954 musical includes some amusing continuity mistakes. For example, during the song 'Wonderful, Wonderful Day' there are live birds flying on the set, but they are clearly confused by the backdrop and several crash into it; and during 'June Bride', the (clearly plastic) icicles on the eaves of the porch are swaying!

7

Hollywood star Elizabeth Taylor has had seven husbands and eight marriages – to Conrad Hilton Jnr (1950-1951), Michael Wilding (1952-1957), Michael Todd (1957-1958), Eddie Fisher (1959-1964), Richard Burton (1964-1974) (1975-1976), John Warner (1976-1982) and Larry Fortensky (1991-1996).

7

Se7en – the 1995 film starring Morgan Freeman and Brad Pitt, tracking a killer who murders to the theme of the 'Seven Deadly Sins'. Grisly dialogue includes the Police Captain saying "Neighbours heard them screaming at each other, like for two hours, and it was nothing new. Then they heard the gun go off, both barrels. Crime of passion". William Somerset replies "Yeah, just look at all the passion on that wall".

7

Ten films have won an amazing seven Academy Awards each,

including *Lawrence of Arabia, Dances With Wolves, Schindler's List* and *Shakespeare in Love.*

7

The seven dwarfs in Walt Disney's 1937 film *Snow White* are Bashful, Doc, Dopey, Grumpy, Happy, Sleepy and Sneezy.

7

The Seven Samurai – the original 1954 Japanese film by Akira Kurosawa that led to Hollywood's 1960 version – *The Magnificent Seven*. The seven actors were Yul Brunner, Steve McQueen, Charles Bronson, Robert Vaughn, Brad Dexter, James Coburn and Horst Buchholz. The Mexican bandit chief was played by Eli Wallach.

7

The Seven Year Itch – 1955 film starring Marilyn Monroe, which includes the then very risqué scene of her skirt blowing up around her waist as she walks over a hot air vent.

8

Butterfield 8 – the title of this 1960 film was based on the New York telephone number of a glamorous model and part-time prostitute, played by Elizabeth Taylor, one of whose lines is "Mama, face it. I was the slut of all time".

8

Title of the 1963 film directed by Frederico Fellini, which won 2 Oscars.

9

The robotic dog K9 (i.e. canine) helped the Time Lord *Dr Who* in the TV series from 1977 until 1981. Originally there were three mechanical versions of the robotic pet. Although he appeared in the 1983 anniversary special, K9 was reunited with the Doctor in the 2005 series.

10

The 1956 biblical epic – *The Ten Commandments* – was the last film of Cecil B. DeMille, the great Hollywood director. Charlton Heston had the lead role as Moses, and it also starred many other greats such as Yul Brynner, Anne Baxter and Edward G.

Robinson. If inflation is taken into account, then this enormously successful blockbuster was the highest US grossing movie of all time – $838,400,000.

10

Romantic comedy film of 1979 based on a scale of attractiveness. It made Dudley Moore and Bo Derrick into superstars, the other star, Julie Andrews, was already a Hollywood icon.

10

Romantic comedy film of 1979 based on a scale of attractiveness. It made Dudley Moore and Bo Derrick into superstars, the other star, Julie Andrews, was already a Hollywood icon.

11

Number of films in which the beautiful Grace Kelly (1929-1982) appeared before she married Prince Rainier of Monaco in 1956.

11

Ocean's 11 – the original 1960 film starred some of the then Hollywood Rat Pack, including Frank Sinatra, Dean Martin, Sammy Davis Jr, Peter Lawford and Joey Bishop. The subsequent remakes of *Ocean's 11* (2001), *Ocean's 12* (2004) and *Ocean's 13* (2007) starred George Clooney, Brad Pitt and Matt Damon amongst others.

12

The Dirty Dozen (1967) included stars such as Ernest Borgnine, Telly Savalas, Charles Bronson, Donald Sutherland and Jim Brown as convicted murderers, recruited for a suicidal mission under the command of Lee Marvin as Major John Reisman. He briefs The Dozen "Shoot any officers you see in there". Victor Franko (John Cassavetes) replies "Who? Ours or theirs?".

12

Twelve Angry Men – 1957 film in which 12 jurors struggle to find the real truth behind an apparently open and shut murder case, summed up by the following quote from one of the jurors, "It's always difficult to keep personal prejudice out of a thing like this. And wherever you run into it, prejudice always obscures the truth. No jury can declare a man guilty unless it's SURE. We nine can't understand how you three are still so sure. Maybe you can tell us."

12

Twelve Monkeys (1995) – the sci-fi film directed by Terry Gilliam and starring Bruce Willis, with the tag line 'The future is history'.

12

Twelve O'clock High – 1949 World War II film featuring Gregory Peck as the tough US Air Force Commander who eventually cracks up.

13

Apollo 13 – the 1995 film starring Tom Hanks about the actual space mission that coined the phrase "Houston, we have a problem".

13

It was originally intended to run TV's *Coronation Street* for only thirteen episodes. It was first broadcast on 9 December 1960, and is still running after 47 years, with nearly 7,000 episodes to date. The working title was *Florizel Street*, until a tea lady at Granada commented that 'Florizel' sounded like a brand of disinfectant. The Street has, on the north side, 7 houses, the pub 'The Rovers Return' and a shop, and on the south side a factory, 2 shop units, a garage and 3 houses.

19

19-19-19 are the rather unusual vital statistics of Olive Oyl – the cartoon character, Popeye's girlfriend.

20

'20th Century Fox', the major film company created in 1935 by merging William Fox's theatre company with Twentieth Century Pictures, started by Darryl Zanuck, Joseph Shenk, Raymond Griffith and William Goetz, whose father-in-law , Louis B. Meyer, provided the finance. It is now part of Rupert Murdoch's News Corporation

21

21 Grams – the film whose title is based on the fact that the body loses this amount in weight at the time of death.

23

23 Railway Cuttings, East Cheam, was comedian Tony Hancock's mythical address in his incredibly popular radio and TV programmes – *Hancock's Half Hour* (1956-1961). Like many a comic genius, Hancock had a depressive side, and ultimately committed suicide in June 1968, leaving a note which said 'Things just went wrong too many times'.

24

There are 24 frames a second on a 35mm moving picture film. Television has 25, to fit in with electricity's 50 cycles a second.

25

Categories of Oscars awarded to the film industry and voted for by the 5,830 members of the US Academy of Motion Picture Arts and Sciences. The Oscars were first introduced in 1929, and the statuettes are 13.5 inches (34 cm) tall and weigh 8.5lbs (3.8 kg). Since 1950, they have been legally encumbered by the requirement that neither winners nor their heirs may sell them without first offering to sell them back to the Academy for $1.

25

Twenty-five per cent of the movie rights of the 1915 blockbuster *Birth of a Nation* was the huge slice that D. W. Griffiths was forced to give the author Thomas Dixon, because he had run out of cash to pay for all the rights. Dixon subsequently made millions.

27

The number of times that Katherine Heigel was a bridesmaid in the film 27 Dresses before she decides to make some changes to her life.

35

Millimetres – the width of traditional movie film stock. Others include 70, 16, and now 8mm and Super8 for home movies.

35

The actual number of minutes, called the 'half', by which time actors have to report for filming or to the theatre stage.

39

The 39 Steps – Alfred Hitchcock's 1935 film starring Robert Donat and Madeleine Carroll. One scene required the two stars to be handcuffed together, and much to their annoyance after it was shot, the practical-joking Hitchcock pretended to have lost the keys for several hours.

50

Twice an Oscar-winning actress, Jodie Foster had amazingly made over 50 movies and TV appearances by the time she went to college. Her first appearance was when she was three, and in her early years she also appeared in several ads.

50

In the 1967 film *Cool Hand Luke* Paul Newman plays a convict who bets that he can eat 50 hardboiled eggs in one hour. He wins – just. However, in real life, the top-ranked professional 'competitive eater', Sonya Thomas, a slim 7-stone American, set the egg-eating record when she ate 65 boiled eggs in a remarkable six minutes and forty seconds.

56

The number of curls famous Hollywood 1930s child star Shirley Temple always had in her hair.

71

Sophia Loren's age in 2006 when she was photographed wearing only lingerie and a bedsheet for the 2007 edition of the famed Pirelli pin-up calendar.

77

77 Sunset Strip – the title of the TV series with Efrem Zimbalist Jr, which ran for 206 hour-long episodes from 1958 to 1964.

101

101 Dalmations – first made as a full length cartoon film in 1961, and then a feature film in 1996, starring Glenn Close as Cruella De Vil supported by Jeff Daniels, Joely Richardson, Joan Plowright and Hugh Laurie – now the star of the TV series *House*.

104

Bullets found in Bonnie and Clyde in the real shoot-out that killed them, and which was subsequently so vividly portrayed in the 1967 film with Faye Dunaway and Warren Beatty as Clyde Barrow who incorrectly predicts "...they ain't gonna catch us. 'Cos I'm even better at runnin' than I am at robbin' banks...".

174

The number of films the authors could find that have a Dracula or vampire theme. (Frankenstein has only clocked up a mere 116). The earliest was *The Secrets of House No 5* (1912), and then *The Vampire* (1913). The first adaptation of *Dracula*, the Bram Stoker novel, was in 1921. But, perhaps the most classic of this genre was Bela Lugosi as *Dracula* (1931). Dracula's more unusual filmic confrontations include sex-symbol *Emanuelle,* comedians Abbott & Costello, cowboy Billy The Kid, comic book hero Batman, and music hall act Old Mother Riley. Some strange horror titles are *Uncle was a Vampire, Bordello of Blood* and *Love at First Bite.*

313

Sweet 16 is the gritty 2002 film directed by Ken Loach which is set in modern Scotland and in which the word 'f*ck' and its variations are used 313 times.

749

As of June 2008 there have been 749 episodes of the world's longest-running sci-fi TV series, *Doctor Who*. The current series is the fourth (1963-1984, 1985, 1986-1989, and 2005 to the present) with a fifth planned for 2010. Ten actors have starred as The Doctor, the first of whom was William Hartnell and the current being David Tennant.

1,200

Lives were saved by German businessman Oskar Schindler when he employed Jewish concentration camp inmates during the Holocaust. Thomas Keneally's biographical book *Schindler's Ark* was made into *Schindler's List* by Steven Spielberg in 1993, and won 7 Oscars.

1984

Film of George Orwell's 1949 book predicting an horrific future, overseen by 'Big Brother'.

2001

2001 A Space Odyssey – the 1968 science fiction film directed by Stanley Kubrick and co-written by him and author Arthur C. Clarke, famous for its visual effects and musical soundtrack.

30,000

Extras in the 1951 film *Quo Vadis*, who used 32,000 costumes. Elizabeth Taylor and Gregory Peck were allocated the starring roles, but they were replaced by Deborah Kerr and Robert Taylor (who had to shave his hairy chest as it was thought too sexy!). Future film star Sophia Loren made her first screen appearance in a bit part as a slave.

150,000

US dollars – the amount for which comic genius Charlie Chaplin (1889-1977) insured his feet. A much-repeated story claims that he once entered a 'Charlie Chaplin look-alike contest' and finished third, although in some versions of the story he came in second. Married four times, he fathered 11 children. Bizarrely, after his death his body was stolen from its grave in Switzerland and only retrieved three months later. It is now encased in concrete.

294,560

Extras used in the film *Gandhi* (1985) for the funeral scene. 200,000 were volunteers and 94,560 were paid a small fee under contract. 11 camera crews filmed 20,000 feet of film for the scene, which was cut to just 125 seconds in the final release.

650,000

US dollars – the amount for which the legs of singing and dancing star Fred Astaire (1899-1987) were insured. The highly erroneous evaluation of Astaire's first screen test was "Can't act. Can't sing. Balding. Can dance a little." Interestingly, Astaire disguised his very large hands by curling his middle two fingers while dancing.

1,000,000

One Million Years BC – the famous 1966 film starring Raquel Welch in a never-to-be-forgotten fur bikini. The bikini, which was invented in 1946, had been banned by Hollywood until Brigitte Bardot made it acceptable in the 1956 French film *And God Created Woman*.

1,000,000

The Million Pound Note was a 1953 film starring Gregory Peck, based on a Mark Twain short story. There are a number of £1,000,000 bank notes in existence, but they are only for internal use in the Bank of England.

1,000,000

One million dollars – the amount that Elizabeth Taylor was paid to play Cleopatra in 1963, becoming the first actress to earn a seven-figure paycheque for a single film.

6,000,000

The Six Million Dollar Man – mid-1970s TV series about a fictional *cyborg* that cost this amount to make, and starring Lee Majors – the husband of Farrah Fawcett Majors.

208,100,000

The greatest total overall cinema attendance of all time was for the film *Gone with the Wind.* Launched in the USA in 1939, then Argentina and Brazil, the film reached the UK in April 1940.

463,000,000

The worldwide TV audience that watched the 1980 cliff-hanging episode of the long-running series, *Dallas*, to find out 'Who shot JR?'.

750,000,000

The estimated worldwide audience for the Royal Wedding of Prince Charles to Lady Diana Spencer on 29 July 1981. There were 39,000,000 viewers in the UK alone. The service in St Paul's was attended by 3,500, and 600,000 people flocked on to the streets of London to watch.

Music Makes the World Go Round

0

Russian composer Pyotr Il'yich Tchaikovsky (1840-1893) never met his patron of 14 years, the wealthy widow Madame Nadezhda von Meck, with whom he corresponded regularly and upon whose support he depended. This suited the morbidly shy homosexual composer, and on the one occasion they were both at the same concert, neither acknowledged the other and each turned away without speaking a word.

0

American singer Mariah Carey is said to be able to cover all the notes from the alto vocal range leading to those of a coloratura soprano, and she has the ability to sing in the whistle register. At one point, The Guinness Book of Records recorded that there was no other singer who could hold a higher note than Carey. The only other mammal capable of a higher register is a dolphin.

1

Only one person accompanied Wolfgang Amadeus Mozart's (1756-1791) coffin to his unmarked grave. Actually this did not reflect his musical standing, but was in line with prevailing Viennese customs of the day. During his life, Mozart composed over 600 works including 21 for stage and opera, 15 Masses, over 50 symphonies, 25 piano and 12 violin concertos and 17 piano sonatas.

1

Pound, the cost of entry in 1970 for the first 1,500 fans to attend the Glastonbury Festival, with free milk thrown in, only veggie food available, and a midnight curfew for amplified music. Nearly forty years later, there are over 300,000 applicants for 100,000 tickets, which cost £155 for the weekend – and there's no free milk.

1

There is just the one continuous groove on a vinyl disc.

1,2,3,4

'*One, two, three o'clock, four o'clock rock*', the opening line of the ground-breaking 1954 number '*Rock Around the Clock*' by Bill Hayley and His Comets, which heralded the arrival of the 'rock 'n roll' era. It was used in the film *Blackboard Jungle*, shot to the top of the charts for eight weeks, and opened the door for stars like Elvis Presley.

2

The classic and enduring song '*Tea for two*' is originally from the 1925 musical *No, No, Nanette*. Lyricist Irving Caesar intended that the original words were to be only temporary, but somehow they survived, and were re-introduced to the public by Doris Day in the 1950 film of the same name.

2

'*Two Little Boys*', sung by Rolf Harris, was the best-selling song of 1969. It was originally recorded by British music hall star Harry Lauder in 1903.

3

Pounds, the total declared value of Johann Sebastian Bach's 6 *Brandenburg Concertos* in 1721. However, these baroque masterpieces were presented to the Margrave of Brandenburg who is thought never to have actually paid for them. The Concertos were initially titled *Six Concerts avec Plusieurs Instruments*, and were only called *The Brandenburg Concertos* by a biographer 150 years later.

3

According to Bob Geldof in 1977 in *Melody Maker* magazine, there are three reasons why people get into 'Rock 'n Roll': "To get laid, to get fame, and to get rich".

3

'*Three Blind Mice*' – a traditional rhyme, and a song, the first written record of which was in 1603 by English composer Thomas Ravenscroft. It possibly refers to the execution of 3 Protestant bishops by Queen Mary.

3

'Three Degrees', soul and disco vocal musical group famous in the 60s. Founded in Philadelphia in 1963, the female group has always consisted of three with new members coming and going. Helen Scott replaced founding member Shirley Porter in 1963 and stayed with the group until 1966; she returned in 1976 and is still with it. All in all, there have been a total of 12 members of the 'Three Degrees'.

3

'*Three Steps to Heaven*' – the posthumous hit for American Eddie Cochran (1938-1960) who died in a road accident in Chippenham while on tour in the UK. David Harman, later known as Dave Dee of the band 'Dave Dee, Dozy, Beaky, Mick & Tich', was a police cadet at the local station, and taught himself to play guitar on Eddie's impounded Gretsch guitar.

4

At the height of their fame, the world-famous group The Beatles – John Lennon, Paul McCartney, George Harrison, and Ringo Starr – were also known as 'The Fab Four'. The name 'Beatles' was suggested by the bassist, Stuart Sutcliffe, who left the band because of mysterious headaches and died a year later. George Harrison replaced him.

4

Antonio Vivaldi's (1678 – 1741) best-known work is his violin concerto, *The Four Seasons*. It was written to go with four sonnets. However, it is not known who the author was – possibly even Vivaldi himself.

4

The number of operas in Richard Wagner's (1813-1883) *Ring Cycle – The Rhinegold*, *The Valkyrie*, *Siegfried* and *Twilight of the Gods* (*Götterdämmering*) which were written between 1848 and 1874, and provide a combined 15 hours of music if performed together.

5

Three days short of his fifth birthday, Wolfgang Amadeus Mozart mastered, in only 30 minutes, his first musical composition. Not long after, he wrote his first, very short, piece of music. In 1762, aged 6, Mozart gave his first public performance, and by the age of 8, had composed his first symphony.

5

'*Take 5*', the cool and emotive Dave Brubeck number that used five beats in a bar.

5

Beethoven's *Fifth Symphony* was the first complete symphony ever to be recorded. During WWII, its distinctive four-note opening, broadcast by the BBC to occupied Europe, became the symbol of victory based on the Morse code letter 'V' (dot-dot-dot-dash).

5

Hours, the remarkably short time it took Richard Rodgers and Oscar Hammerstein to write both the music and the words of the 1943 hit *Oklahoma!*, winning them a special Pulitzer Prize.

5

Mozart's piano had only five octaves, which is why he never wrote in very high or low notes.

5

The age at which Charlie Chaplin suddenly took over his mother's music hall role when she fell ill – and money showered on to the stage. Years later, when he returned to England, such was Chaplin's international fame that he received 73,000 letters in just two days.

5

The Jackson 5, one of many pop groups using numbers in their names, such as The Three Degrees, The Four Tops, U2, and UB40.

5-4-3-2-1

Theme tune composed by subsequent rock star Manfred Mann and his band, The Manfreds, for ITV's iconic pop music programme *Ready Steady Go* (1963-1966). '*5-4-3-2-1*' rose to No 5 in the UK charts.

9

The age at which Russian musical child prodigy Sergei Prokofiev

(1891-1953) composed his opera *Giant.*

10

It took the Beatles just ten days in 1964 to sell one million copies of '*I want to hold your hand*' as 'Beatlemania' swept through America.

10

When Gustav Mahler died in 1911, he was working on his tenth symphony, now called *The Unfinished Symphony*, and regarded as one his greatest works.

12

Practically all of the classic jazz blues have twelve bars, of which the greatest exponent is probably W.C.Handy.

14

Librettist W.S. Gilbert (1836 – 1911) and composer Arthur Sullivan (1842 – 1900) co-operated on fourteen comic operas between 1871 and 1896. They were brought together by producer Richard D'Oyly Carte, who built the Savoy Theatre in 1881 to stage their works – by his own D'Oyly Carte Opera Company – after which they became known as the 'Savoy Operas'.

14

The age at which Duke Ellington (1899-1974) composed his first song – '*Soda Fountain Rag*'. He went on to a career of international repute, including making jazz widely acceptable to the general public through broadcasting from New York's famous Cotton Club, and which spanned more than four decades.

15

Fifteen minutes going home in a cab was all Irving Berlin needed to write the music and words of '*Anything you can do, I can do better*', to add into the Broadway musical '*Annie Get Your Gun*'.

16

'*Sixteen tons and what do you get, another day older and deeper in debt*' was written by Merle Travis, and made into a hit by 'Tennessee' Ernie Ford.

16

...in fact, sixteen seems to be a popular age in the world of songs. '*16 Candles*' was recorded by the Crests in 1959, and then by several other artists including Roy Orbison.'*Sweet Sixteen*' by Hilary Duff in 2003 celebrates the freedom of being a sixteen-year old girl and was used as the theme tune for the TV show *My Super Sweet 16*. Billy Idol had a song of the same name in 1987. Neil Sedaka sang '*Happy Birthday Sweet Sixteen*' (1961), and Chuck Berry '*Sweet Little Sixteen*' (1958). American punk band FallOut Boy recorded '*A Little Less Sixteen Candles, A Little More "Touch Me"* ' in 2006.

17

Minutes, the length of Maurice Ravel's repetitive orchestral piece *Bolero*. One of its most public airings was to accompany ice-skaters Torvill and Dean to their Olympic gold medal-winning performance in the 1984 Winter Olympics, at which they achieved a perfect score.

19

Rock legend has it that Rolling Stones Mick Jagger and Keith Richards first collaborated in 1964, when Andrew Loog Oldham locked them in a kitchen saying, "Don't come out without a song". Their charting 1966 song '*19th Nervous Breakdown*' was one such – possibly about Jagger's then girlfriend, model Chrissie Shrimpton. Over 90 Jagger/Richards songs have been released, and when they co-produce records, the two work under the pseudonym 'The Glimmer Twins'.

24

'*Twenty-Four Hours to Tulsa*' – Gene Pitney's hit song, one of many using American place names, such as – '*Wichita Lineman*', '*I Wish I was in Peoria*', '*The Man from Laramie*', '*The Girl from Ipanema*', '*St. Louis Blues*', '*Rock Island Line*', and rather rudely, '*Lubbock, Texas, in our rear view mirror*'.

26

Country and Western singers originally had to perform for 26 days a year to maintain their membership of the legendary venue – the Grand 'Ole 'Opry, Nashville, Tennessee. As an artist's appearance fee was only $44, and most had many more profitable engagements, this has been considerably reduced – but they are still expected to show commitment by frequent performances.

27

The incredible number of curtain calls for Maria Callas on the opening night of *Tosca* at the Royal Opera House, in 1964. The ovation lasted no less than 40 minutes.

28

Twenty Eighth Street, New York – the original, noisy 'Tin Pan Alley' music-writing street.

29

The track (platform) from which Glenn Miller's '*Chattanooga Choo Choo*' leaves Pennsylvania Station in New York.

40

Songs that have included the word 'Broadway', the former cart track across New York that became the centre for music and theatre.

50

Curtis Jackson (b 1975) is better known as the rap singer '50 Cent'. He had a troubled upbringing on the streets of New York, and in 2000 he survived being shot nine times. His music reflects this background – such as his 2003 charting album '*Get Rich or Die Tryin*'.

51

'*Reel of the 51st*' – the Scottish dance invented in 1940 by officers of the 51st Highland Division while they were prisoners of war.

60s

The sixties were a seminal period in the age of music and emerging personal freedoms. As old rockers say of the self-indulgent and often drug-induced decade, "If you can remember the sixties, then you weren't there".

66

'*(Get your kicks on) Route 66*' is a 1946 rhythm and blues classic song, which has been covered by various artists such as Nat King

Cole and the Rolling Stones. Route 66 was already a literary legend (immortalised in John Steinbeck's *Grapes of Wrath*) and eventually featured in a TV series. The song's lyrics record a journey along America's first trans-continental all-weather highway – 2,400 miles long and crossing 8 states – 'Well it goes from St. Louis down to Missouri / Oklahoma city looks oh so pretty / You'll see Amarillo and Gallup, New Mexico / Flagstaff, Arizona don't forget Winona / Kingman, Barstow, San Bernadino. / Would you get hip to this kindly tip / And go take that California trip / Get your kicks on Route 66'. It has now been bypassed by the Interstate highways, but its motels and gas stations are being preserved for posterity.

70

The longest time an album has topped the charts in the UK was 70 weeks – for the soundtrack of the musical *South Pacific*. This is followed by The Beatles' '*Please Please Me*', which was at the top for 30 weeks. The longest stay in the UK charts was 278 weeks between 1970 and 1975 for Simon and Garfunkel's '*Bridge Over Troubled Waters*'.

88

Over the years the name of the musical instrument 'gravicembalo col piano e forte' (harpsichord with pedals soft and large) has eventually been shortened to 'piano'. A standard piano has 88 keys, and 220 strings – tightened to a tension of 160 lbs.

97

'*The Wreck of Old 97*', the ballad about a 1903 train crash, recorded by Vernon Dalhart in 1924, and the first country record that sold more than a million copies. Also recorded by Johnny Cash, Woody Guthrie, Hank Williams III, Boxcar Willie and The Seekers.

126

The seminal 60s and 70s rock band 'The Who' are believed to hold the record for the loudest ever rock concert at 126 decibels (measured at 32 metres from the loudspeakers) at Charlton Athletic Football Ground in 1976.

200

The number of songs created to celebrate Charles Lindbergh's solo flight across the Atlantic in 1926, in addition to the 'Lindy Hop' dance. After Lindberg's solo non-stop flight from New York to Paris, in which he 'hopped' over the Atlantic, Shorty George Snowden was dancing in a marathon contest at the Manhattan Casino in Harlem when a reporter asked him what dance he was doing. The headlines in the newspapers had stated 'Lindy Hops the Atlantic', so he told the reporter, "I'm doing the Lindy Hop."

235

Number in Basin Street in New Orleans' Storyville red light district, the home of Lulu White's Mahogany Hall, in 1900 the finest brothel in America – hence the '*Mahogany Hall Stomp*' by Louis Armstrong, 1929 (he probably delivered coal there as a kid).

281 F

The cover of the Beatles' eleventh album, *Abbey Road*, took just ten minutes to shoot. The result is an iconic image of the group walking across a 'zebra crossing' with a Volkswagen 'Beetle' co-incidentally, but appropriately, parked behind. The car, which belonged to people living in a flat opposite, had the registration 281 F, and, as the album became famous, the number plate was continuously stolen. Eventually the car was sold in 1986 for $23,000, and is now on display in Germany.

536

The great American singer songwriter Bob Dylan (born Robert Zimmerman in 1941) has, since 1962, recorded 32 studio albums, 13 live albums, 13 compilation albums and 2 with the super group 'The Travelling Wilburys'. Dylan has probably written 536 songs, but the number is difficult to specify as not all are copyrighted. As one anonymous internet blogger parodied Dylan's 'Blowin' in the Wind' – 'How many songs must Bob Dylan write / Before you call him a poet? / Yes 'n' how many lyrics must one man sing / Before everyone will know it? / Yes 'n' how many years must one man tour / Just so he can show it? / The answer my friend is ...'

1,003

The number of lovers accredited to Mozart's 1787 fictional operatic hero, *Don Giovanni*, was perhaps modelled on the great rake, Casanova.

1742

The first ever performance of Handel's '*Messiah*' was given on 13 April 1742 at Dublin's Fishamble Street Music Hall by the combined choirs of St Patrick's Cathedral and Christ Church, with Handel himself leading the orchestra on the harpsichord. When King George II first heard the *Messiah* and it reached the '*Hallelujah Chorus*' movement, he rose to his feet. Royal protocol of the day decreed that if the monarch stood up, so did everybody else. Thus, it became traditional for audiences to stand as the first notes of the '*Hallelujah Chorus*' sound out.

1812

Pyotr Tchaikovsky's *1812 Overture* celebrates Russia's epic defence against Napoleon's invading army at the Battle of Borodino.

4,381

Apparently well over four thousand women gave birth at Rod Stewart's concert on Copacabana Beach, Rio de Janeiro in 1994. They were part of the largest audience in history, no less than three and a half million.

6-5000

'*Pennsylvania 6-5000*' was not only the Glenn Miller Orchestra's 1940 swing hit record, but also the oldest continuous New York telephone number – from 1919 it has been that of the Hotel Pennsylvania. Many other famous bands played at the hotel, including the Dorsey Brothers, Count Basie and Duke Ellington.

300,000

The number of UK album sales required for the artist(s) to be awarded a 'Platinum Disc'. In the US this figure is 1,000,000 – but there is also a 'Gold Disc' (UK = 100,000 and US = 500.000). But, the US also has a 'Diamond Disc' which requires sales of 10,000,000 – this applies to both singles and albums. The first 'Gold Record' was awarded to legendary bandleader Glenn Miller as a publicity stunt in 1942, for selling 1.2 million copies of his song *'Chattanooga Choo Choo'.*

28,000,000

The best-selling album of all time is by The Eagles – *'Their Greatest Hits (1971-1975)'* with sales of 28 million – although, ironically, it did not include perhaps their best-known track '*Hotel California*', which was not released until 1976. Next is Michael Jackson's 1982 album '*Thriller*', and third is Pink Floyd's 1979 album *'The Wall'*.

30,000,000

The second best-selling record ever, with 30 million sales, was Bing Crosby's '*White Christmas*', made famous by the eponymous 1954 film.

37,000,000

Elton John's record '*Candle in the Wind*', (or '*Goodbye England's Rose*') is the world's biggest-selling record by a recording artist, with 37 million copies – this 1997 version was in memory of (Princess) Diana, and sung by him at her funeral. The original version by Elton John, in 1973, was a tribute to American movie star, Marilyn Monroe.

1,000,000,000

The number of Elvis Presley records sold in 50 years. In the USA, 149 of his songs entered the *Billboard's* Hot 100 Chart, and 114 made it to the top 40 – and of these, 40 made it into the Top Ten.

1,500,000,000

The worldwide audience of 'Live Aid 'on 13 July 1985 – the first global charity rock concert that lasted 16 hours, was watched in 100 countries, and is heralded as the day that 'Rock 'n Roll' changed the world. Each band was allowed a 17-minute set, and Phil Collins managed to perform in both London, and then (by flying Concorde to the USA) in Philadelphia, USA. Live Aid raised £110 million.

-275

The coldest recorded temperature is 1 billionth of a degree above absolute zero, -275 degrees C (minus 460F), achieved by MIT scientists when cooling sodium in 2003.

0

Tortoises do not have any teeth; however, they do have a sharp beak which is used to bite food.

0

An earthworm has no eyes.

0

Dalmatian puppies are born white and spotless, and only develop spots after ten to fourteen days.

0

New Zealand is one of only two countries in the world with no indigenous snakes – the other is Ireland.

0

There is no water stored in a camel's hump, which is in fact fatty tissue that helps sustain a camel over the 5 to 7 days that it can go without food or water. The tissue converts to water when it comes into contact with oxygen from air carried in the camel's blood stream. A camel can take in 20 to 25 gallons (91 to 113l) in a single drink, and then lose up to 25% of its weight before cardiac damage (most other mammals can only lose 3 to 5%).

.03

The speed in mph (0.048 kph) at which a snail crawls.

.118

Inches (3 mm) is the thickness of the ozone layer if evenly concentrated in one layer. In reality, it is spread out through the stratosphere, and even small amounts, diluted with other gases, play a key role in soaking up the sun's ultraviolet radiation.

1

One square mile (2.6 sq km) of land contains more insects than the earth's total human population.

1

The female hyena's genitalia are disguised as an apparent scrotum and a prominently displayed false penis, which is in fact a clitoris, through which runs its vagina.

1

A bedbug can survive one year without eating.

1

A cubic mile (4.16 ckm) of fog contains less than one gallon (4.5l) of water.

1

The unicorn was a mythical animal that had only one horn, and by tradition, cloven feet. Its primary appearance is as a heraldic device. Nobody can be sure of its origins; perhaps the rhinoceros being the only extant animal on the planet with one central horn (or now extinct variations) is a possible source. Another is the narwhal – a whale whose home is in the Arctic. Of the narwhal's two front teeth, the right one stays embedded in the skull and normally grows to about 1 foot (0.3 m) in length; however the left one grows out straight with an anti-clockwise spiral, to a length of about 8 feet (2.4 m).

1.5

Gallons (6.8 l) – total amount of water an elephant can store in its trunk. An elephant drinks 18-36 gallons (82-163 l) a day.

2

Hours – the time it takes bacteria to reach danger levels in cooked food left at 70° F.

2.2

The production of just 2.2 lbs (1 kg) of beef emits the same amount of the greenhouse gas, carbon dioxide, as the average European car driving 155 miles (250 km).

2.2
The seed pods of one cocoa tree produce 2.2 lbs (1 kg) of chocolate. The cocoa pods contain 25 to 40 seeds, and weigh 7 to 28 ounces (200-800 g). They take 5 to 6 months to ripen and each tree is harvested twice a year.

2.5
Ounces – the weight of the brain of a 6.4 ton (6,502 kg) *Stegosaurus*, probably making it the world's stupidest creature.

3
A chimpanzee can mate very, very quickly – in 3 seconds. Female chimps tend to make more noise during mating when there are aggressive males around, so they can draw attention to themselves and have as many partners as possible. However, they are quieter when other females are around because they do not want them to be jealous and break up the union.

4
Four-leafed clover. Deemed lucky, because in botany most plants have uneven numbers. One old saying on the luck of the clover: 'One leaf for fame/One leaf for wealth/And one leaf for a faithful lover/And one leaf to bring glorious health/Are all in a 4-leaf clover'.

5
Years – the age that a filly becomes a mare, and a colt becomes a stallion – unless of course he's been gelded, in which case he remains forever a gelding.

5
Porpoises and dolphins can stay under water for about 5 minutes between breaths. Their cousins, the killer whales, normally achieve 6 minutes, but one has been recorded at managing 17. Hippopotamuses and crocodiles normally dive for the same time, but crocodiles have been known to achieve more than 2 hours, by slowing their metabolism right down.

5
The total worldwide tonnage of diamonds mined annually.

5

Yards that a giant tortoise can crawl in one minute.

6

With six per cent of the world's population, the United States uses 25% of its resources and creates 25% of its pollution.

6

Six seconds, the estimated time it took dinosaur *Compsognathus* to run 100 metres, a speed of 40 mph. This dinosaur was the size of a small cat, weighed 6.6 lbs (3kg) and lived 150 million years ago. This compares with the fastest biped on earth today, the ostrich, which can manage a top speed of 34.5mph.

6

The central crystals forming a snow crystal have 6 sides. Snowflakes are not frozen raindrops. Sometimes raindrops do freeze as they fall, but this is called 'sleet'. Snow crystals form when water vapour condenses directly into ice, which happens in the clouds. It takes a snowflake 9 minutes to fall from 1,000 feet (304 m).

6

The earth's gravity is 6 times greater than that of the moon.

7

Mph (11.2 kph) – the speed at which an average raindrop falls.

7

The number of bones in a giraffe's neck – the same number as those in a human neck. However, at 6 feet, that of the giraffe is somewhat more extended. Also, a giraffe's tongue is 3 feet (91.44 cm) long.

7

The common ladybird has seven spots.

7

The average amount of semen in a male elephant's ejaculation is between 7 and 10 fluid ounces (200 to 300 ml) – enough to fill a Coke can. An elephant can mate up to 5 times a day over 3 days, and each testicle can weigh as much as 5.5 lbs (2.5 kg).

8

The barnacle's sedentary lifestyle means that it has had to evolve some unusual characteristics. The male barnacle can extend its penis up to a remarkable eight times the length of its body, and also adapt the shape to suit its maritime environment – in gentle water, a thinner penis is better for maximum reach, while in rough seas a stouter one is more effective.

8

Inches (20.3 cm) – length of a newborn crocodile. After being hatched from the egg, infants are carried to safety in the mouths, or throat pouches, of their mothers.

8

Legs on a lobster.

10

Feet (3 m) – the length to which a rabbit's front teeth would grow it they were not worn down by eating.

10

Inches (25.4 cm) of snow is equivalent to one inch (2.4 cm) of rain.

10

The population of New Zealand is outnumbered by sheep by a ratio of 10 to each person. This used to be much higher in 1982 with 22 sheep per person, but sheep numbers have declined while the human population has grown. Similarly, Australia now has only 4.55 per person, partly due to drought devastating the flocks.

12

An ostrich egg is 24 times the size of a hen's egg, and weighs up to 3 lbs. You could feed 12 people with an omelette made from one egg, or make 100 meringues or 32 soufflés.

12

The famous game bird, the Red Grouse, is only found in the UK. The grouse shooting season starts on the 12 August, known universally as the 'Glorious Twelfth', and continues until 10 December.

14.2

Hands (58 inches/145 cm) – the height at its withers above which a pony is defined as a horse.

18.75

The largest hailstone ever recorded fell in June 2003 in Aurora, Nebraska. It was 7 inches (17.7 cm) in diameter, and had a circumference of 18.75 inches (47.62 cm).

20

An oak tree does not produce acorns until it is at least 20 years old.

22

Hours a day that the Koala Bear sleeps, making the sloth, needing only 20 hours, seem rather hyperactive!

24

Hours, the short lifespan of a mayfly. Juvenile mayflies can live for up to two years in streams and rivers. However, once they emerge in the open air as adults, they live for just a few hours or a day. This part of their lifespan is incredibly short, and basically dedicated to one thing before they die – breeding.

27

Days, the time it takes duck eggs to hatch, while a chicken egg takes around 21 days.

33

The largest Giant Squid actually caught (New Zealand 2007) was 33 feet long (10 m) – including mantle and its two tentacles (this squid also had 8 arms) and weighed 1,091 lbs (495 kg). However, rumours abound of even larger squids of at least 59 feet (17.9 m) in length. Giant squids are preyed upon by sperm whales, and, as they live very deep (over 1,000 feet/304 m) in the ocean, their eyes can be well over one foot (.3 m) in diameter.

33

The Asiatic Reticulated Python is thought to be the world's longest snake – at 33 feet (10 m). South American anacondas are almost as long, but considerably thicker. Rather disgustingly, at 33 feet, a human tapeworm can be about the same length.

35
The rate in inches (90 cm) that a bamboo can grow in a day.

35.3
Cubic feet (1 cubic m) of water weigh 2,204 lbs (1,000 kg), or just short of a ton.

40s
The 'Roaring Forties', and 'Furious Fifties' – intense low pressure systems that form in the Southern Ocean from cold air from Antarctica meeting warmer air from the ocean, resulting in some of the most ferocious winds on the planet.

42
The number of teeth in an adult dog – a human has 32.

43
Mph, (69.2 kph) the speed of an Afghan hound, nicknamed 'the greyhound in pyjamas'. In contrast, greyhounds reach 35 to 40 mph (56 to 64 kph) in races.

50
Although scientific estimates vary considerably, it is thought that fifty plant and animal species become extinct each day through pollution and land exploitation.

74
Mph, (111 kph) is the minimum speed at which a wind becomes a hurricane.

75
Per cent of the earth's surface is covered by water – less than 3% of which is fresh. Of this, 70% is locked in Polar icecaps, 29% stored as groundwater in underground aquifers and less than 1% in lakes and streams.

80
Metres – if all the ice in glaciers and ice sheets melted, this is by how much the sea level would rise, flooding huge areas of the earth.

80
Times its own height that a flea can jump, the equivalent of a human leaping over a building 440 feet high. It can also jump 150 times its length, the equivalent of a human jumping 250 yards.

91
Feet (27 m) – the highest recorded wave – in the Gulf of Mexico during Hurricane Ivan – on 16 September 2004.

100
Times a second, the frequency that lightning strikes somewhere on earth. Every hour, 1,800 thunderstorms are taking place – amounting to 16 million storms a year.

100
Yards (91m) a mole can burrow in a night.

100
Years that a sea anemone can live.

127
Average number of species of trees and plants found in one square mile of rainforest. Rainforests are self-watering, with canopy trees releasing water vapour into the atmosphere – as much as 200 gallons (760 l) a year. This moisture forms a cloud, which, when not falling as rain, keeps the forest humid and warm.

136
Miles per second (220 kps) – the speed at which our solar system orbits the Milky Way Galaxy.

180
In the 1800s, the world's longest creature, a ribbon worm measuring 180 feet (58.4 m), was washed up on a Scottish shore. They regularly grow to 160 feet (50m).

190
Beats per second of a housefly's wings.

200
Theoretical human deaths that a Black Mamba snake could achieve with the venom of one bite.

231
Mph, (371 kph) - the highest wind speed recorded, on Mount Washington in 1934.

280
Parts per million of the greenhouse gas, carbon dioxide, in the atmosphere in the year 1000. This gas is mainly caused by burning fossil fuels. This level remained fairly constant until the early 1800s, when, with the arrival of 'Industrial Revolution', it escalated violently. Now the figure has risen to 380 parts per million.

302
The length of Ireland at its longest is 302 miles (485 km); at its widest 189 miles (304 km); and it has a coastline of 3,500 miles (5,631 km). The country is divided into four historical provinces – Connacht (5 counties in the west); Munster (6 counties in the South); Leinster (12 counties in the East); and Ulster (9 counties in the north).

323
The egg not only has an effective aerodynamic shape, it also has an inherently strong structure. The record for throwing an uncooked egg in its shell without breaking is held by Texan Johnie Dell Foley, who in 1978 achieved the distance of 323 feet 2 inches (98.5 m). It was caught undamaged by his cousin.

350
The Wellingtonia tree (*Sequoiadendron giganteum*) is the world's largest recorded living thing – past or present. It has been measured up to 350 feet tall (107 m), with a base diameter of 35 feet (10.7 m), and the greatest reported age of 4,000 years.

400
Watts per kilogram (2.2.lbs) of mass – energy generated by a fast flying grouse – *Lapogus lapogus scotticus* – while 80 watts per kilogram of mass is the energy generated by a professional sprint cyclist.

405

The Quahog Clam, caught off the coast of Iceland in 2006, is at 405 years old the longest-living animal ever known. By the time it was recognised in a laboratory to be so old, it had unfortunately died! In comparison, mere youngsters in the animal kingdom are the Galapagos tortoise achieving 176 years, the sturgeon 154 years, the bowhead whale 130 and an elephant clocking up a more modest 78 years.

600

The pressure applied by the teeth of a Great White shark is 600 pounds per square inch – enough to shear through steel. Lions apply the same pressure, but hyenas and snapping turtles can achieve 1,000. Humans only bite at 120 psi (827 kPa), but one crocodile bite has been recorded at 6,000 psi (42,368 kPa).

600

Years, the time since the year 1130 it took four carefully guarded rabbits to finally escape into the wild in Britain and spread into the countryside. In spite of the 1950s ravages of myxamotosis, the average population of rabbits in winter is now 40 million, but this can rise to 300 million in summer.

645

The gestation period for elephants is 645 days, rhinos 480, giraffes 425, goats and sheep 150, dogs 61, cats 62, rabbits 33 and mice 19. Humans need 265 days, which, co-incidentally, is similar to orang-utans at 260 days, gorillas 257, and chimps 237.

750

The number of legs on the world's 'leggiest' creature – the female millipede (*Illacme plenipes*), each only 1.5 inches (38mm) long. The male is smaller and has up to 400 legs – two of which have modified into sex organs.

1,000

In one year, one female and one male rat will have produced 1,000 progeny. This might explain why the UK population of brown rats, (*Rattus norvegicus*), is estimated at 80 million, with an appetite for at least 210 tons of food a year.

1,000

The canary's heart beats 1,000 times a minute.

1,000

Years or more – the age of the oldest tree in the UK – the Bowthorpe Oak in Lincolnshire. With a 40 feet (12 m) circumference and a hollow trunk, 39 people can stand inside it, and at least 13 could sit down for a meal.

1,359

The Dead Sea surface is 1,359 feet (414 m) below sea level. Two thousand years ago the water level was much higher at 1,188 feet (363 m), but lack of water from the Jordan River and evaporation means that the level is dropping all the time.

1,590

There are only 1,590 Chinese Giant Pandas still living in the wild. Their existence is endangered by a number of factors - their natural habitat is increasingly being encroached upon by man; a slow rate of reproduction; the scarcity of the arrow bamboo, their favourite food; and, more recently, the devastating 2008 earthquake.

2,000

The 'silk' produced in a spider's web is the strongest of all natural and man-made fibres, and has a greater tensile strength than steel. Some spiders can 'spin' up to a continuous length of 2,000 feet (609m). As an indication of its strength, it would have to be 50 miles (80.5 m) long before it broke under its own weight. Spiders' webs have many different functions – to trap prey, as a safety line, a cocoon for eggs, a home, and even to hold on to their partners!

4,500

Number of crocuses that it takes to make 1 ounce (28.3 g) of saffron.

4,767

Years – the age of the world's oldest tree – the Bristlecone Pine in east California, USA. Nicknamed 'Methuselah', its exact location is a closely-guarded secret.

6,500

Feet (1,900 m). The thickness of ice that covers Antarctica, and accounts for nearly 85% of the world's ice.

7,270

Miles – the non-stop flight from Alaska to New Zealand by 'Miranda', a Bartailed Godwit, which beat her wings continuously for 8 days, neither eating nor sleeping. Fitted with a satellite tracking device, she became a public heroine in New Zealand and bells were rung in Christchurch Cathedral on her arrival.

7,927

The block of gold in cubic metres (279,902 cubic feet) which could be produced from the 153,000 tons of gold ever mined – this equates approximately to the size of three London town houses.

10,000

It is estimated that there are ten thousand privately owned tigers in the USA, dwarfing the 200 or so kept by zoos. Conservationists estimate that even in the wild in India, the population of tigers might have sunk in seven years from 5,000 to as low as 1,300.

33,000

Feet (10 km) – the height to which cumulonimbus clouds may rise from their base near the earth's surface.

46,000

Miles per annum – the distance a Sooty Tern flies to migrate, making it the longest annual migration of any species. It can cover 550 miles (885 km) per day in a figure-of-eight flight from New Zealand to the north Pacific via Japan, Alaska, and California. These birds dive to an average of 46 feet (14 m) below the surface to catch fish, but can reach 225 feet (68 m).

46,000

Lbs (20,700 kg) of earth mined to produce a half-carat diamond.

100,000-300,00

Scoville Units – this is the heat specification of the hottest ever pepper, the 'habeñero'; on the other hand, the 'jalapeño' rates only 2,500-5,000 on the Scoville scale. Christopher Columbus first discovered chilli peppers in the West Indies on his first voyage. Now amongst the top 75 plants consumed in the world, there are between 2,000 and 3,000 different types of chilli.

830,000

The number of people killed in the worst recorded earthquake, in Shaanxi, China, in 1556.

1,000,000

The number of phytoplankton (tiny one-celled plants that drift in ocean currents) that can be contained in a teaspoon.

1,500,000

Scientists differ widely as to exactly how many species there are on earth – numbers quoted vary from 3m to 112m. However, whatever the number, humans have named only 1.5 million. The estimated number of species of living mammals is 4,000. This compares with 8,600 species of birds and 23,000 of fish. All of which is dwarfed by 20,000 different kinds of ground beetle, of which the UK alone has 340 types, and 800,000 identified species of insects. Scientists believe that there could be a further million or more insect species still to be identified.

12,000,000

The number of nematodes in a square metre of soil. Nematodes are tiny wormlike creatures, of which there are 80,000 types, 15,000 of them parasitic. They were first recognised as an agricultural pest when microscopes became powerful enough to see them, and they are thought to destroy up to 10% of the world's crops.

33,000,000

The theoretical progeny of two rabbits after 2 years.

166,000,000

Although the very strange-looking Australian Duck-billed Platypus diverged from other mammals 166 million years ago, it still shares 82% of its genes with many different relatives, including humans and mice. Uniquely, the male has poison spurs on its hind legs – the genes that produce this venom are similar to those found in snakes. Another reptilian characteristic is that the Platypus lays eggs.

220,000,000

A dog's sense of smell is 44 times more sensitive than that of a human – no doubt due to the 220 million smell-sensitive cells in the nose. However, it is estimated that dogs can smell 1 million times better than humans because they can detect multiple layers of chemicals. Today, it is even thought that they can 'smell out' cancer and other diseases.

250,000,000

The length of time that the cockroach as a species has existed on earth.

4,500,000,000

Years ago that Earth started forming. It was much later, 3.8 billion years ago, that the earth started cooling, causing water droplets, then rain, then the sea and subsequently life.

10,000,000,000

There are an estimated 10 billion langoustines (aka Dublin Bay prawns) in British waters – 75% of which are found off Scottish coasts. The breaded tails of langoustines are found on menus as scampi.

10,000,000,000

The number of bacteria in a gram of soil.

326,000,000,000,000

This a figure about which we have difficulty being precise! There are around 326 trillion million gallons of water on earth.

0

At 5.29 a.m. on 16 July 1945 the first atomic bomb was exploded in New Mexico. The very secret 'Manhattan Project' at Los Alamos developed the atomic bomb with a view to bringing World War II to an end before the Germans developed their own version. 'Ground Zero', the spot where a bomb detonates, was named by 'the father of the atomic bomb', Robert Oppenheimer, Scientific Director of the project. The site of the Twin Towers, destroyed by the 9/11 attack on the World Trade Center, was also given the name 'Ground Zero'.

0

General Sir George Everest (1799-1866) never once saw the mountain that bears his name. The Great Trigonometrical Survey of India was begun at Cape Comorin in 1806, and taken over by a new Surveyor General, Everest, in 1823. As it worked north, the logistics of this remarkable survey required 4 elephants from which to watch for tigers, 30 horses for officers, 42 camels for supplies and 700 labourers. Due to ill health Everest had to hand over the Survey to General Sir Andrew Waugh, who in 1856 named 'Peak XV', the world's highest mountain, 'Everest' out of respect for his predecessor.

0

Sadly, no hard evidence of the 'Loch Ness Monster' has ever been found despite many 'authentic' sightings. The earliest record of a monster was in 565 AD during the life of St Columba, and local reports continued over the centuries. In 1871 there was a possibly dubious report by a D. McKenzie, and modern sightings began in earnest in 1933 with a number of eye witness accounts of a long-necked monster seen on land as well as in the water of the loch.

0

The Greenwich Observatory lies on 0 degrees longitude, and since 1884 has been the place from which all the world's time zones are calculated – known as Greenwich Mean (or Meridian) Time (GMT) or Zulu Time, based on the solar time at the

Observatory. Standardisation of time in Britain was started in 1847 because of railway timetables, and was legally adopted throughout the country in 1880. The Observatory's latitude is 51° 28′ 38″ N (North of the Equator).

0

The number of people who have seen 'Big Ben' from Parliament Square. 'Big Ben' is in fact the bass bell, hidden from view. People call St. Stephen's Tower 'Big Ben' in error.

0

The only major landmass to have no glacial ice is Australia.

0

There are no American states that do not have a vineyard – even Alaska has three wineries, Mississippi two and Hawaii one.

1/9th

The geographical extent of Russia's landmass in relation to the world's land area. Russia is 70 times as big as the UK, but has only 2.3 times the population.

1

Mile (1.6 km), the depth of the Grand Canyon. It is also, incredibly, 217 miles (349 km) long.

1

According to the Useless Information Society, St John's Wood is the one and only London tube station whose letters do not appear in the word MACKEREL. Surely a perfect example of the justification of the Society's existence!

1

In 1900, Britain was the premier nation for wealth, military power, world business, education, currency and standard of living. How things have changed.

1

One o'clock in the afternoon, the time that the 'Timeball' at Greenwich descended (and still does), so that ships departing down the Thames could set their chronometers. At midday they would have been too busy operating their sextants, taking a fix on the position of the sun.

1

The remarkable and unique address – No 1 London – belongs to Apsley House on Hyde Park Corner, which was the home of the victor of the Battle of Waterloo, the first Duke of Wellington. Originally built in the 1770s for Lord Apsley, it was bought in 1807 by the Marquis of Wellesley, but in 1817 financial difficulties forced him to sell it to his younger brother – the Duke.

1.25

The 'Rowley Mile' Race Course at Newmarket has a straight that is in fact 1 mile and 2 furlongs (2 km) long. One of two Newmarket racecourses, it is named after King Charles II's favourite horse – Old Rowley - a nickname that was later given to the monarch himself.

1.34

The eventual length of Southend Pier was 1.34 miles (2.1 km). First opened in 1889, it was extended over the years to accommodate the tide that recedes for over a mile.

2

There are two Ballingarry villages in County Tipperary, making it confusing if you were trying to find the Ballingarry site of the 'Battle of Widow McCormack's Cabbage Garden' during the 'Young Ireland' revolt of 1848.

2

Names given to US Government buildings because of their visual appearance – the White House and the Pentagon.

2

The United States and Russia are separated by just over two miles (4 km) in the Bering Strait – between Russia's island of Big Diomede and the USA's island of Little Diomede.

3

It took just three hours of the conference at a villa in Berlin's lakeside Wannsee, in January 1942, for a small group of 15 senior

Nazis to plan the process to exterminate the 11 million Jews in Europe, achieving, horrifically, 6 million deaths.

3

Number 3, Abbey Road, London, was the address of the recording studios that the Beatles made famous. At their first session, producer George Martin gave the Beatles a lecture about the importance of using better quality amplifiers, ending his diatribe with the query, "Look, I've laid into you for quite a time, and you haven't responded. Is there anything you don't like?" A long pause followed, with shuffling of the Beatles' feet. George Harrison stared at the producer and said, "Yeah, I don't like your tie!" And so began the placement of Abbey Road Studios as a shrine in the eyes of Beatle fans.

3.97

The degree of leaning of the belltower at Pisa. This means that the 7th and highest cornice protrudes 4.5 metres beyond the lowest. Construction of the Leaning Tower of Pisa was started in 1173, an edifice 60 metres high and weighing 14,500 tons.

4

Standard time divisions in the USA.

5

Miles – the distance around Hendon representing Britain's first 'Green Belt' – an area protected from building development.

5

The 'Cinque Ports' (Sandwich, Dover, Hythe, Romney and Hastings) were established in 1155 by Royal Charter to protect England's south-east coast.

5

The Hagia Sophia in Istanbul took only 5 years to build between 532 and 537 AD by Emperor Justinian 1. This is remarkable when compared with other great Christian cathedrals such as St Paul's Cathedral in London which took 33 years, and Notre Dame in Paris, 87. The Hagia Sophia was Christian until 1453 when it became a mosque, and was then secularised as a museum in 1934.

6

The number of people who died in the Great Fire of London in 1666. This was a miracle considering that 13,000 houses, 400 streets and 87 churches were engulfed by flames.

7

The surprisingly tiny number of prisoners 'liberated' by the French revolutionary mob from the Bastille prison on 14 July 1789 (four forgers, two lunatics and one sex offender).

7

The number of oaks on the Tonbridge Road after which Sevenoaks was named. Sadly, 4 of these blew down in the great hurricane of 1987.

007

The extensive facilities at Pinewood Film Studios include 18 stages, the largest of which is called 'The Albert R. Broccoli 007'. With an usable area of 59,000 square feet (5,481 sqm) it is the largest stage in Europe.

9

Nine Elms is an area of London – south of the river bordered by Battersea and Vauxhall (where the first Vauxhall car was constructed on the site of what is now Sainsbury's petrol station). In Victorian times, it had a major London train terminus until the line was extended to Waterloo. The station was knocked down and the New Covent Garden flower market relocated there in 1974.

10

10 Downing Street, London, is the official London residence of the British First Lord of the Treasury and not, as normally thought, that of the Prime Minister. However, since 1902 the two positions have been held by the same person, so it has become associated exclusively with the Prime Minister. Unusually, No 10's door can be opened only from the inside. Downing Street was originally built in the 1680s. Number 11 Downing Street, is, by tradition, the home of the Chancellor of the Exchequer who is the Second Lord of the Treasury. Number 12 used to be the Chief Whip's residence, but now houses the PM's press offices, while the Whips

have moved to number 9.

10

Miles in a 12-hour day – the record speed for laying railway track, achieved in 1869 by 10 Irishmen of the Central Pacific. They used 3,250 rails and 25,000 wooden sleepers, and each man lifted 11 tons (11,176 k) of rail.

14

People killed when a B-25 bomber crashed in fog into New York's Empire State Building in 1945 – a tiny number compared with the almost 3,000 killed in the World Trade Center in 9/11.

20

Letters in the longest place name in the Republic of Ireland – Muckanaghederdauhlia, in Co. Galway.

31

The number of times the island of Tobago has changed hands between Spain, Britain, France and Holland.

32

There is a piece of stone from every one of Ireland's 32 counties in the construction (1899-1904) of St Anne's Cathedral in Donegall Street, Belfast.

34

Miles (54.7 km) – the shortest intercontinental flight in the world – Gibraltar (Europe) to Tangier (Africa). However, the shortest distance between two continents is across the Bosphorus in Turkey, where less than two miles (3 km) separate Europe and Asia.

42

42nd Street, New York, centre of the Broadway theatre district and one the many places made immortal in song; *'59th Street Bridge', '12th Street Rag'* and *'Slaughter on 10th Avenue'*.

47

Czars are buried within the Kremlin complex.

53

Years after the invention of the telephone, a line was installed for the first time in the Oval Office in the White House, for President Herbert Hoover in 1929.

54

No. 254, West 54th Street, home from 1977-79 to the notorious 'Studio 54' nightclub, where guests were chosen on the pavement according to whether they were 'happening enough'.

61

Per cent of all the world's imported cut flowers are sold through the Dutch flower markets, and Europe consumes over 50% of the world's flowers.

66

Days – the amazingly short time it took to repair the freeway system and its bridges in Los Angeles after the earthquake of 1994, 74 days ahead of schedule. The contractors gained $200,000 for each day they were early, and would have lost $205,000 for each day late.

66

Route 66 – the first 'all-weather' road, stretching 2,400 miles across America. It transformed the country and became the subject of innumerable films and songs, for example '*Get your kicks on Route 66*'.

66

The very high percentage of world's copper that was mined in Cornwall in the nineteenth century, although Cornwall was more famous for its tin.

67

Staircases in the magnificent 17th century Palace of Versailles – one of the largest castles in the world. It has more than 2,000 windows, 700 rooms, 1,250 fireplaces, 1,400 fountains and 1,800 acres of park. The splendid Hall of Mirrors is 250 feet long and would have been lit by over 3,000 candles.

73.5

Miles (118 km), the length of Hadrian's Wall across the North of England, built by the Romans to keep out the Pictish tribes of Scotland.

75

The last two figures of French car number plates denoting the Paris area, soon sadly to be phased out.

80

Feet – average depth of the Persian Gulf. The English Channel, also shallow, is only 150 feet (45 m) between Dover and Calais but averages 390 feet (120 m).

100

A total of 100 bridges cross the River Thames over its 215 mile (346 km) length – starting at Thameshead in the Cotswolds and ending at Dartford. Within the London area there are 34 bridges, starting with Walton Bridge and ending with the four-laned Queen Elizabeth II Bridge with a main span of 1,476 feet (450 m); this, along with back spans, making an overall length of 9,423 feet (2,872 m).

100

Feet. (30.4 m) – the length of the Ironbridge at Coalbrookedale, Shropshire, built in 1779 celebrating Abraham Darby's invention of cast iron produced from coke rather than charcoal – and thus giving birth to the world's iron and steel industries.

100

Megatons, the force of the Krakatoa volcano explosion in 1883, heard 2,000 miles (3,220 km) away.

100

Miles of underground roads which serve the huge salt mines in Winsford, Cheshire. They are high enough to take a double-decker bus and lit by street lamps.

138

Number of The Spanish Steps in Rome.

162

The Burj al Dubai tower in Dubai has 162 floors and is currently the tallest building in the world, rising to 2,684 feet (818 m).

167

Letters in the full poetic name of Bangkok – *Krungthepmahanakhon Amornrattanakosin Mahintharayutthaya Mahadilokphop Noppharat Ratchathaniburirom Udomratchaniwetmahasathan Amonphiman Awatansathit Sakkathattiyawitsanukamprasit* – which translates as 'The city of angels, the great city, the residence of the Emerald Buddha, the impregnable city (of Ayutthaya) of God Indra, the grand capital of the world endowed with nine precious gems, the happy city, abounding in an enormous Royal Palace that resembles the heavenly abode where reigns the reincarnated God, a city given by Indra and built by Vishnukarn'.

221

221B, Baker Street, the famous mythical home between 1881 and 1904 of fictional detective Sherlock Holmes. In reality, Baker Street has never had a house with this number – 100 being the highest ever allocated.

246

"How can you govern a country with 246 varieties of cheese?" – famously said by Charles de Gaulle of France, the country of which he was the President.

250

The number of people who have fallen off the Leaning Tower of Pisa.

279

As statues go, that of Mother Russia at Mamaev Kurgan, the memorial to the dead at the WWII battle of Stalingrad, is probably the world's tallest at 279 feet (85 m). She is 171 feet (52 m) high and her sword a further 108 feet (33 m). The 'Christ the Redeemer' on the 2,300 feet (70 m) high mountain above Rio de Janeiro is 120 feet (36 m), and the outstretched arms spread 100 feet (30m). Antony Gormley's 'Angel of the North' is a modest 66 feet (20m) tall, but with an impressive 178 feet (54 m) wingspan.

355

Yards (325 m), the length of the streets of Pamplona which annually witness the 'Running of the Bulls', an event which has resulted in 14 deaths and 200 injuries.

423

The circumference in yards (376 m) of the Great Court at Trinity College, Cambridge, which students try to run around in 43 seconds – the time it takes the college clock to strike 12 noon – made famous by the film *Chariots of Fire* (although this particular scene was filmed at Eton). Incredibly difficult to do, and only achieved by Lord Burghley in 1927 and Sam Dobin in 2007. Sebastian Coe, with a time of 42.53 seconds, didn't quite make it, as the clock had been rewound the day before, making the chimes ring slightly quicker than normal.

530

Miles (853 kilometres) – the length of the shelving in the US Library of Congress in Washington.

821

The lowest population of any country in Europe is that of the Vatican City with just 821 inhabitants in its 44 hectares. There are 558 official citizens who include the Pope and 58 cardinals; the remainder serve in roles such as the Papal Diplomatic Service and the Swiss Guard. About 3,000 additional local employees are Italian and other nationalities.

900

The population of San Francisco in 1848. With the 'Gold Rush' the next year, this jumped to 100,000.

999

Emergency telephone number in the UK and Ireland – the US equivalent is 911.

1,050

Sandstone blocks made up the Abu Simbal temples, transported 200 feet (61 m) higher to avoid flooding by the Aswan High Dam.

1,150

Feet (350 m) below sea level – the depth of the cavernous salt mine at Carrickfergus in Northern Ireland that provides most of the anti-ice salt for the UK's roads.

1212

The old telephone number, Whitehall 1212, of the Metropolitan Police Headquarters at New Scotland Yard. It is still contained in their contact numbers by many UK police forces.

1,500

Miles (2,415 km) – the longest hedge in history, built in 1845 by the British across India to stop salt smuggling. Part of a 2,500 mile (4,025 km) guarded custom line, the Great Hedge was the second longest structure in the world after the Great Wall of China.

1600

Pennsylvania Avenue – the address of The White House, home to United States Presidents.

2000

BC – the founding date of Damascus, Syria, the world's oldest still-inhabited city.

2,575

Over one quarter of the 10,000 glass panels on Boston's prestigious John Hancock Centre fell off between 1971 and 1973. The danger of the 500 pound panels falling required surrounding streets to be closed, and while the problem was being solved, the temporary replacements ensured it was soon called the 'Plywood Palace'. The cause was never publicly revealed, and eventually the contractor replaced all the glass at a cost of some $7 million.

2648

The house number at Grand Boulevard, Detroit, where Berry Gordon started his world-famous record company, Motown, spawning stars like 'The Supremes' and 'The Jackson Five'.

3,000

Miles (4,830 km) – the length of the highway that the Incas built down South America.

3200

BC – the date when the Newgrange Passage Tomb in Ireland was constructed, making it 600 years older than the Giza Pyramids and 1,000 years older than Stonehenge.

3,500

Miles – the length of the Silk Road, the legendary trading route from the east in China, to Europe via the Middle East.

3764

The number in Elvis Presley Boulevard, Memphis, Tennessee, where the King's 'Graceland' mansion has become the second most popular tourist address in America, after the White House.

3,946

Miles (6,353 km), the length of the Great Wall of China. Built to defend China's northern borders, its first construction was in the 7th century BC and continued in various forms until the 1600s, but the main part of the building took shape around 200 BC. It probably cost the lives of 2 to 3 million slaves to build, and would have been guarded by about 1 million troops. It is a myth that it can be seen from the moon.

4,145

Miles (6,673 km) – the length of the Nile River, 138 miles (222 km) longer than the Amazon.

4,440

Number of statues in Milan Cathedral.

7,107

The number of islands that make up the Philippines.

9,000

Years – the length of the lease on the original Guinness Brewery at St. James's Gate in Dublin, first granted at a rent of £45 per annum in 1759. The brewery has subsequently grown tenfold in size.

11,072.76

Miles (17,820 km) – the total length of the coast of Great Britain.

13,659

The number of wooden piles driven into the clay to support the foundations of Amsterdam.

20,000

Volunteer skilled workers are now known to have built the Great Pyramid of Giza, as opposed to the 100,000 slaves who were once thought to have done so.

27,133

Square miles (70,275 sq km) – the total area of the Republic of Ireland, consisting of a land area of 26,580 square miles (68,842 sq km) and 553 square miles (1,433 sq km) of water. England is 50,352 square miles (130,410 sq km).

29,796

Mauna Kea, the dormant volcanic mountain on the island of Hawaii, could be considered the world's highest mountain. It rises 29,796 feet (9,082 m) from its base on the adjacent ocean floor. This is just over 750 feet (228 m) higher than Mount Everest, whose height is measured from sea level. Mauna Kea's height above sea level is 13,796 feet (4,205 m), with 16,000 feet (4,877 m) below water level.

42,649

People to the square mile in Monaco, the most crowded country on earth. The next is Singapore with a population of 18,476 per square mile, which is followed by Malta with 3,205.

200,000

People who fled their homes during the Three Mile Island nuclear accident in Pennsylvania.

250,000

Rome's ancient Circus Maximus was Rome's enormous site for horse and chariot racing. Its foundations were originally started in 6th century BC, and at its greatest it was 2,000 feet (610 m) long, 500 feet (152 m) wide and had a capacity of 250,000 – a quarter of Rome's population. In comparison, the Colosseum amphitheatre, inaugurated in 80AD, had a capacity of 45,000 in its three-tiered 615 feet (187 m) by 510 feet (155 m) elliptical format.

293,000

Panes of glass (over 1 million square feet – 92,903 sq m) in the incredible 1851 building, the Crystal Palace. Other remarkable numbers: it was designed in just 10 days by Sir Joseph Paxton and erected in only 8 months by 2,000 workmen to a budget of £79,800. Its 990,000 square feet (91,974 sq m) housed the 13,000 exhibits of the Great Exhibition which were seen by 6,200,000 visitors. The whole building was moved to Sydenham Hill, South London in 1854, and was destroyed by fire in 1936.

1,000,000

In 800 AD, Baghdad was the largest city on earth with a population of one million. It was a seat of great learning and a major centre of the Muslim world. In 1258, it was sacked by the Mongols and many of the citizens were killed. Its population today is about 7 million, making it the second largest city in the Arab world – Cairo being the largest.

2,500,000

The iconic French structure, the Eiffel Tower, is held together by 2.5 million rivets. Built in 1889, it is 984 feet (300 m) high plus 79 feet (24 m) antenna) but this can vary by as much as 6 inches (15 cm) in the summer heat. It requires 50-60 tons (50,800-60,960 kg) of paint every six to seven years. Although it is the most paid-for attraction in the world (with over 200 million visitors since being built) not everybody approved of its design. French novelist Guy de Maupassant was a vociferous opponent, but when asked why he lunched in its restaurant every day, he replied that "This is the one place from which I cannot see the Tower!"

3,000,000

Three million tourists visit Hershey, Pennsylvania, each year – the home of the American confectionery giant, and the ultimate company town, with its Chocolate Avenue and Cocoa Inn.

5,000,000

Five million people a year visit the pilgrim site of Lourdes in France, to obtain a miraculous cure. As a result, its airport is one of the busiest in France.

7,200,000

London has the largest population of any UK city (as defined by a large town with a cathedral) with 7.2 million people. St David's has the smallest with only 2,000. Of the capital cities, Edinburgh has a population of 430,000, Cardiff 292,000, Belfast 276,000 and Dublin 860,000.

10,000,000

New York's 84-storey Empire State Building consists of 10 million bricks, and was built in just a year and 45 days at the height of the Depression, using the most sophisticated building and materials-delivery systems of the day. It was opened in May 1931 by President Hoover, but it was a whole decade before it was fully let, and was dubbed 'The Empty State Building'.

20,000,000

The number of those who passed through the Immigration Station of Ellis Island between 1892 and 1954 on their way to find the 'American Dream'. 120 million Americans are descended from them today.

649,600,000

Ounces – the largest amount of gold ever stored at Fort Knox, – just after the attack on Pearl Harbor in 1941. The current amount stored in the US Depository is 147 million ounces, worth $6,206,3400,000 at book value.

0

Even though the South Africans have won two Rugby World Cups (1995 & 2007), they have scored absolutely no tries in either of these finals. By contrast, the English team has scored only one try in its three World Cup finals – 1991, 2003 (winners) and 2007.

1

The top motor racing category is Formula 1, or Grand Prix racing. Over the years, the size of the engine and the weight and dimensions of the cars have varied, but today 2.4 litre naturally-aspirated engines provide an amazing 720 break horse power. Previously only produced by turbo charging, such outputs are only possible because the engines can rev at 19,000 rpm due to advanced metallurgy and pneumatic valves.

1

One hour, the time a boxer is allowed to try to reduce his weight if he fails the first weigh-in.

1

Only one cricketer has ever hit a six over the Pavilion at Lords, and that was in 1899 by the remarkable, but now long-forgotten, Albert Trott (1873-1914). Trott played for Australia in 1894-95, and today still holds the highest batting average (102.5) in all the Australia v England series. When Trott was surprisingly omitted from the next Australian team, he sailed to England, played for Middlesex and subsequently for England on a tour of South Africa. Despite his incredible record with bat and ball, Trott sadly committed suicide in 1914, having written his will on a laundry ticket – leaving his wardrobe and £4 to his landlady.

1.58

Ping pong balls used in international competitions have a diameter of 1.58 inches (40mm) and weigh .09 ounces (2.7 g). A couple of fun facts: first, 343 balls would fit into a cubic foot, and second, they can stand 60 psi of pressure, so could be sunk to a depth of 90 feet (27.4 m) of water without collapsing.

2

British boxer Henry Cooper twice fought Mohammed Ali – perhaps the greatest boxer of all time. Towards the end of the fourth round of their first match (1963) Cooper punched Ali (then called Cassius Clay) so hard that he went down and could hardly get up. However, Cooper was unable to follow up and knock out his opponent, because the bell went for the end of the round. Ali's trainer, Angelo Dundee, used the fact that his man had a torn glove (the hole is thought to have been enlarged by the trainer) to delay the beginning of the 5th round, thus giving Ali more time to recover. He went on to beat Cooper in this fight, and again in 1966. However, Ali never forgot the effect of *'Enry's 'Ammer',* as Cooper's very effective left hook was known.

3

All modern British thoroughbred horses trace their origins back to three stallions imported into England from the Middle East – *The Byerley Turk* (1680s), *The Darley Arabian* (1704) and *The Godolphin Arabian* (1729). One study of equine Y chromosomes indicates that 95% of all thoroughbreds can be traced back to *The Darley Arabian.*

3

Balls used in billiards – two cue balls, one of which is white and the other white with a spot, and one object ball which is normally red. The dominant material from the 17th to the 20th centuries was ivory but as this was very expensive, it was replaced by modern materials Bakelite and plastic. Snooker uses 22 balls – 1 white cue ball, 15 red and 6 of different colours.

3

On the 2 April 1977, gelding *Red Rum* galloped into history by winning the Grand National for the third time. His other two wins were in 1973 and 1974 (he was second in 1975 and again in 1976). *Red Rum's* total winnings were £114,000.

3

Since the modern Olympic Games started in 1894, they have been cancelled only three times – due to war – in 1916, 1940 and 1944.

4

Points deducted in show-jumping by a refusal, and with two refusals resulting in elimination. Four penalty points are also deducted for a knockdown, but only if it changes the height of the obstacle – thus, if a middle or lower rail is knocked down with the rider and horse clearing the top, then no points are deducted.

4

There are four main individual Gaelic sports administered by the Gaelic Athletic Association. Hurling and Gaelic Football are the main ones, but the Association is also responsible for Rounders (very similar to the UK and American versions) and Gaelic Handball, which is not unlike Fives.

4

There are four players in a polo team, and each will be graded by a handicap from -2 to 10 (being the highest and best). The total handicap of the team is added together and the team with the lowest is awarded the difference in goals at the beginning of the match.

5

In 1909, keen boxing fan and Patron of the National Sporting Club, Hugh Lowther, the 5th Earl of Lonsdale, instituted boxing's oldest championship award – the Lonsdale Belt. It is still awarded for all weights by the British Boxing Board of Control, and much prized by its recipients. The first winner was lightweight Freddie Wells. To keep the Belt, a boxer must first win it and then successfully defend it twice. The first to achieve this feat was heavyweight Henry Cooper in the 1960s.

5

The number of players in each team playing on court in basketball matches. The modern game was developed by a Dr. James Naismith in Springfield, Massachusetts in 1891, when he was looking for an indoor game to keep his students occupied. Initially, he nailed a peach basket 10 feet up a wall into which players aimed the ball. But, as it had a bottom, the balls were difficult and slow to retrieve. Once this was cut out, the balls could fall through – the start of the modern fast-scoring game.

6

Each ice hockey team will have 6 players on the ice at any one time. The etymology of the word 'hockey' is uncertain, but could be from the Old French word *hoquet* meaning shepherd's crook – which may be the origin of the hockey stick. Ice hockey is mostly associated with Canada, Russia, Sweden and the USA. However, a total of 64 countries are now playing members of the International Hockey Federation.

6

A six iron was the golf club that astronaut Alan Shepard used to hit 2 golf balls on the surface of the moon in 1971. Thanks to the lack of gravity, he was able to claim jokingly that his second shot went "for miles and miles and miles".

7

There are seven players in a water polo team – six offence and defence players, and a goalkeeper. Teams also have three reserves.

8

Seconds – a key number in rodeos, which originated to display the five essential skills needed by cowboys – team roping, saddle 'bronc' (unbroken horse) riding, calf roping, bareback riding and bull riding. For the last two events, the rider must stay on the bucking animal for 8 seconds using only one hand, and is also judged on his technique, control and the strength of the horse's or bull's bucks.

8

The normal number of greyhounds in a race. Greyhound racing is the second most popular spectator sport in the UK – after soccer.

8

The number of lanes in an Olympic swimming pool. Each lane is 8 feet (2.5 m) wide, and the overall pool length is 164 feet (50 m).

10

Ten minutes was the length of time, in May 2007, that football manager, Leroy Resenior, was in charge of Torquay United. Immediately after being appointed, he was told that in fact the club was being sold.

10

The most usual weight of a boxing glove in a professional match is 10 ounces (283 g). Obviously, the heavier the glove the more protection there is, so training and sparring boxing gloves tend to be 12-16 ounces (340-454 g).

10

There are ten ways a batsman can be out in cricket – 1) Bowled 2) Caught 3) Leg-before-wicket (LBW) 4) Stumped 5) Run out 6) Timed out (after 3 minutes) 7) Obstructing the field 8) Handling the ball 9) Hit wicket 10) Double hit.

11

Number of players per side actually on the field at any one time in American football. However, teams are very much larger (up to 46) as players have specific roles such as defence and offence, and can be substituted as required. The pitch measures 120 (109.7 m) by 53 yards (48.5 m) wide; the goalposts are 100 yards (91.4 m) apart; and behind each is the 10 yard (9.1 m) scoring area – the 'end zone'.

12

Question: what is the minimum number of balls that have to cross the net to complete a set of tennis? It's a lot fewer than you might think. There are a maximum of eight serves per game and six games make a set, so one of the obvious answers might be 8 x 6 = 48. Wrong. The answer is – one player serves 4 winning aces with his first serve for each of his three games, and then his opponent hits all his serves into the net, so none actually makes it over for his three service games. Thus, only 12 balls have crossed the net before the six-game set is completed.

12

The dimensions of full-sized snooker and billiards tables are 12 feet (3.65 m) by 6 feet (1.82 m).

13

Feet (4 m) – the distance a tug-of-war team has to pull the opposing team to win. Each team consists of 8, and the hemp rope is normally 110 to 118 feet (33.5 to 36 m) in length.

15

Players in a Gaelic football team. Matches normally last 60 minutes, although senior games are 70 minutes long and are played with a round ball slightly heavier than a soccer ball. The players move the ball up the field with a combination of carrying, dropping the ball on the toe and then kicking it back into hand, kicking and passing – all with the object of getting the ball through the opposition's goal.

16

'Sweet 16' refers to the last 16 teams in the American College Basketball Championships, which narrow down to the 'Elite 8' and then the 'Final 4'.

17

Seventeen riders took part in the first of what was to become the Grand National steeplechase in 1839. One rider, Captain Becher, did not finish, falling into the brook that subsequently bears his name, and afterwards exclaiming "I had not realised how dreadful water tastes without whisky in it".

18

Over the ages there have been quite a number of regional variations of 'dartboard', and some can still be found in Staffordshire, Manchester, and Yorkshire. However, it was in the mid 20th century that the layout was standardised with a width of 18 inches (45.7cm), and 20 being the highest-scoring segment (although landing the dart in the narrow inner ring scores treble the segment value). The outer small circle in the centre scores 25, and the inner smaller circle 50. The board is hung so that the bull's eye is 5 feet 8 inches (172.5 cm) off the ground, and the distance from the throwing line to the board is 7 feet 9 inches (237 cm).

18

There are many differences between Association Football, Rugby Football and the 'Australian Rules' game. In the Australian version there are 18 players per team, and the ball can be advanced by kicking, handling and running, and, unlike most other sports, there is no offside rule. Its origins are obscure, but it is thought to have originated in the country's goldfields in the mid 1850s.

21.45

Hours and minutes – the time it took Captain Matthew Webb to swim the English Channel in 1875, the first man to do so. With tides and currents, he is thought to have swum 39 miles (62.7 km). He died in July 1883, trying to swim the rapids of the Niagara Falls, and is buried there in the local Oakwood cemetery.

24

Feet (7.3 m) – the distance apart of the inner edges of Association Football goalposts. The lower edge of the horizontal bar must be 8 feet (2.43 m) above the ground. The net is optional. For international matches, the length of the rectangular pitch must be in the range of 110-120 yards (100.5-109.7 m), and the width 70-80 yards (64–63 m).

24

The first Le Mans 24 Hour endurance motor race was on 26 and 27 May 1923 for sports cars 'complete with a passenger and tools'. It has been run every year since, except in 1936 (economic depression), and 1940-1948 (WWII and its after-effects). The 2008 winner averaged 134.4 mph (216.3 kph), in spite of the fact that the Mulsanne Straight now has chicanes to stop the cars reaching 250mph (402 kph).

24

World Heavyweight boxing champion, Sonny Liston (1932-1970), was thought to have 24 brothers and sisters.

30

In the 4.5 mile (7.24 km) long steeplechase, the Grand National, there are a total of 30 fences, and all but two (the Chair and the Water) are jumped twice. At 5 feet 2 inches (5.12 m), the Chair is the highest fence in the race, and the race finishes with a 494 yard (452 m) run-in.

37.5

Japan's national sport, Sumo wrestling, is one in which size and power are distinct advantages in the short 4-minute rounds that take place in the 15 feet (4.55 m) sacred *doyho* or ring. Thus, one top-class wrestler, the Hawaiian born Konishiki, was pre-eminent – weighing in at 37.5 stone (238 k). Not surprisingly, he was also known as the 'Dump Truck'! The traditional Sumo loin cloth, or *mawashi*, wrapped around the wrestler's waist, is often as long as 30 feet (9 m).

40

The Grand National steeplechase is limited to 40 runners, and the greatest number of finishers was 23 in 1984, with the fewest being two in 1928.

46

Even today, Argentinian Juan Manuel Fangio (1911-1995) is considered by many to be the greatest racing driver of all time. His remarkable record in 52 races was as winner in 24 of them, with 35 podium finishes. Fangio was World Champion 5 times (1951, 1954, 1955, 1956, & 1957), and was 46 when he won his last Grand Prix in a Maserati, consequently becoming World Champion.

49

The speed in mph (79 kph) at which Cristiano Ronaldo of Manchester United managed to kick a football to score one of the best penalty goals ever seen in the English Football League, on January 30 2008 in a match against Portsmouth. This is relatively slow, as some kicks are thought to be as fast as 70-80 mph (112-128 kph).

55

Strokes – the lowest recorded score on an 18-hole golf course was by PGA and Champions Tour player, Homero Blancas, in 1962 on the 5,000 yard (4,572 m) course at Longview, Texas. He had 13 birdies, one eagle and used only 20 putts.

59

Minutes – the time it takes to complete a 15-round boxing match if it goes the distance (15 x three minute rounds, and 14 x one minute breaks).

70

At 70, Sir Eyre Massey Shaw is the oldest contestant to win an Olympic Gold Medal – for yachting in the 1900 Paris Games. In fact, his medal was made of silver as the winners' medals were not made of gold that year.

96

Britain's worst sporting disaster saw 96 Liverpool soccer fans crushed to death due to an overcrowded stand at Hillsborough Stadium in Sheffield, during the FA Cup semi-final between Liverpool and Nottingham Forest on 15 April 1989. As a result, new stand layouts were introduced – mostly with seats. Four years earlier, 39 supporters had died at the Heysel Stadium in Belgium when trouble broke out between Juventus and Liverpool fans. As a result, there was a blanket ban for five years on English clubs playing in European competitions.

115

Seconds – this was how long English football head coach Steve McClaren spent at the press conference after his team only just managed to defeat Andorra in March 2007 – his comment being, "Gentlemen, if you want to write whatever you want to write, you can write it because that is all I am going to say. Thank you". The press did just that. When he was subsequently sacked after the team failed in November to qualify for Euro 2008, McClaren received a £2,500,000 pay-off. During his 18-match period of management, he capped 45 different players. In comparison, when Jose Mourinho left Chelsea he received £12,000,000.

141

Mph – the first speed record on Bonneville Salt Flats, recorded in 1914. The current world speed record by the British Thrust SSC (acronym for Super Sonic Car) stands at 763 mph (1,228 kph).

170

Inches – the normal length of a boxer's protective wrap for an average-to-large hand. This is wrapped around a boxer's hands before he puts on his gloves, to protect the bones and tendons and to support the thumb and wrist.

200

Miles per hour – the speed a shuttlecock can reach in the game of Badminton – the game created by the bored children of the Duke of Beaufort in 1863, and named after their home at Badminton House. This compares with a squash ball reaching 151 mph (242 kph) and a tennis ball at 138 mph (222 kph).

336

The normal number of dimples on a golf ball, although they vary from 300 to 450. The reason a golf ball has dimples is to reduce the drag factor. Dimples haven't always been round – there have also been rectangles, squares and hexagons.

448

Bret Melson, a student in Hawaii, set a new world record in February 2007 – for the longest hole-in-one ever scored on the 448-yard (409 m) par 4 at the Ko'olau Golf Club in Oahu. Previously, the longest straight shot hole-in-one in golf history was hit by Robert Mitera in 1965 at the Miracle Hills Golf Club in Omaha, Nebraska, where he used his driver on the 10th hole from 444 yards (405 m). However, in July 2002 at the Green Valley Ranch Golf Club, Michael J. Crean of Denver, Colorado, drove off the tee on the 517 yard (517 m) 9th hole. The ball landed on the green and rolled into the cup, but as the hole was a dogleg and his drive had cut the corner, this was not recognised as golf's longest hole-in-one.

496

Singles victories by English tennis player, Tim Henman, during his 14-year professional tennis career until he retired in 2007. Henman also endured 274 losses, but he was in 6 Grand Slam semi-finals.

643

The number of times the ball passed over the net in the longest-ever rally in a competitive tennis match. This was between Jean Hepner and Vicky Nelson in 1984 at Richmond, Virginia. The match took 6 hours 22 minutes, concluding with a 1 hour 47 minute tie break, and one point alone took an amazing 29 minutes.

1,000

The Mille Miglia was a sportscar race on 1,000 miles (1,609 km) of open roads with little crowd control, and was run annually in Italy from 1927 to 1957. It was won in 1955 by Stirling Moss in a Mercedes SLR averaging 99 mph, with his navigator Dennis Jenkinson shouting the route details from a long scroll. Two years later, the Marquis de Portago's Ferrari crashed killing 12 spectators, and as a result, this very dangerous race was stopped for ever.

1,007

The longest golf hole in the world is the 1,007 yards (1,101 m) at Chocolay Downs, Marquette, Michigan. It is a 6 par.

1,576

There is an annual running race up the 1,576 steps of New York's Empire State Building. The record time from the ground up the 86 flights to the top is 9 minutes 33 seconds.

1829

This was the year when Cambridge student, Charles Merivale, issued a challenge to his Oxford friend Charles Worsdworth (nephew of the poet William) for a boat race. This took place on 12 March at Henley-on-Thames, and was watched by 20,000 people with Oxford the clear winner. In 1845 the race moved to its current course starting in Putney. The 1877 Oxford and Cambridge Boat Race resulted in a dead-heat, the only time this has ever happened. It is reported that the finish judge actually called it "a dead-heat to Oxford by five feet".

1863

The year that the Football Association banned the handling of balls by players other than the goalkeeper.

1878

When Manchester United was first formed in 1878, its name was Newton Heath LRW (Lancashire Railway Workers). In 1880 it was called just plain Newton Heath, and then in 1902 it became Manchester United.

1890

That most American of games – baseball – has in fact been played in the UK since 1890, and at various times been very popular. The football element of Derby County Cricket Club moved to a new ground in 1895, which it then shared with the baseball players. The interest in baseball waned while that in football thrived, but Derby County's permanent home is still called The 'Baseball Ground'.

10,500

The number of athletes expected to have participated in the

2008 Beijing Olympics over 16 days at 37 different venues within China. The Olympic programme consisted of 28 sports split into 302 individual categories - 165 men's events, 127 woman's and 10 mixed. Contestants came from 204 different countries, and, while the host nation China's team had 639 members and the USA 595, some sent just single entrants. Over 40 countries did not participate. They tended to be the smaller nations or island states such as The Vatican City, Gibraltar, St. Helena, Christmas Island, Montserrat and Greenland. The worldwide audience for the opening ceremony was about 4 billion.

52,000

The total number of balls used each year during the two weeks of the Wimbledon Lawn Tennis Championships. About 20,000 of these are used during practice and qualifying, and then the rest in matches, where there are new balls after every 7 games. Each two ounce (56.7 g) ball is tested for weight and compression, so that when it is dropped from a height of 100 inches (254 cm) on to concrete, it must bounce between 53 and 58 inches (134-147 cm).

60,000

The average salary of a county cricketer in 1998 was in the region of £23,000 – today this is £60,000. If the player's team wins competitions, then his salary might be boosted to £150,000. A National Test player can earn £400,000 with winning bonuses. However, top class players in the new international Twenty/20 competitions can earn as much as £2 million.

125,000

There are 125,000 active football teams in England which have a total of about 7 million players (3m adults and 4m children) and which participate in 1,700 different leagues. All this requires at least 26,000 referees to police the games.

150,000

Pounds, the final cost in 1907 of the motor racing circuit at Brooklands (£11 million today). This expense nearly broke its owner, Hugh Locke King, who had thought that it might cost the same as creating a golf course.

205,000

Spectators in 1950 for the World Cup match between Brazil and Uruguay.

400,000

Having burned 123,900 calories, pedalled over 812,000 times at an average speed of 24 mph (38 kph), worn out three cycle chains and sipped from 42,000 water bottles over the three weeks and 2,241 miles of the annual endurance cycle race, the Tour de France, the winner receives £400,000.

492,308

The amount in £s that footballer David Beckham is thought to earn a week as a result of his move to US football club Los Angeles Galaxy. His basic pay is a relatively modest $5.5 million, but this is supplemented by a percentage of the tickets, sales, sponsorship and other royalties. In comparison, the average salary for professional players in the US is $115,432 (£59,195). Such was the interest in Beckham's move to the US that, at his first press conference, the accredited media numbered over 700.

15,000,000

Spectators who line the route of the 2,174 miles (3,500 km) Tour de France cycle race, the world's largest live sports event. The 198 riders are supported and accompanied by 4,500 people, including 14 medical staff in 1,500 vehicles. Following the race are 2,300 accredited journalists and 1,200 photographers, cameramen and TV directors who represent 78 channels from 170 countries with a worldwide audience of 2,000 million.

0

Even if you shouted in space and somebody was right next to you, they would not hear anything at all. On Earth sound, like light and heat, travels by making molecules vibrate, but in deep space there are no molecules.

0 and 1

The binary system, or 'base-2 number system', uses the combinations of two digits – usually 0 and 1 – to make up the internal language for all modern computers. Although partly developed in 11th century China, Francis Bacon investigated a system in 1605 in which letters of the alphabet could be reduced to binary sequences. Over the following years various other mathematicians developed the theory, until 1937 when George Stibitz completed a computer based on strings of binary relays – 'The Model K' ('K' for Kitchen – where he assembled it).

1/3rd

The gravitational force on the surface of Mars is only about one third as strong as that on the surface of the Earth.

1

"One small step for man, one giant step for mankind" – the words of astronaut Neil Armstrong – the first man to walk on the moon on 20 July, 1969.

1

The Mohs Scale is the universally used guide to the hardness of minerals. Talc, which has number of industrial uses over and above talcum powder, is at the bottom of the scale with 1, and a diamond is the hardest with a rating of 10. A harder mineral can scratch a softer mineral, but a softer one cannot mark a harder one – however hard you try.

1

As an example of the vastness of space – if the earth's solar system were reduced to the size of one US quarter coin (25c), then the Milky Way in proportion would be the size of North America.

1

Mach 1, the speed of sound at sea level, 763 mph (1,229 kph), named after the Austrian physicist and philosopher Ernst Mach (1838 – 1916).

1

The thickness of optic fibre used in telecommunications is normally one millimetre. The fibres are normally made out of glass or plastic fibre, and are more effective than metal wire because they are immune to electromagnetic interference, and signals travel along them with less loss.

1

The energy required to raise the temperature of one pound of water from 59.5° F to 60.5°, defined as a 'British Thermal Unit'. BTUs are the units of energy used in the power, steam, generation, heating and air-conditioning industries.

1

Inches (2.5cm), the diameter of a stroke of lightning. The 'return stroke' travels at 62 million mph (100 million kph), and its temperature is 33 thousand degrees C.

1

Venus is the one and only planet that rotates clockwise.

1

'J' is the one letter of the alphabet that does not appear in the 'The Periodic Table of the Chemical Elements', the framework used in Chemistry to classify and display the properties of all the many different chemicals.

1

To have the number '1' as the fastest of anything may seem surprising, but in the very competitive race to develop the world's fastest supercomputers, the latest models being developed will have a peak performance of one 'petaflop' (one quadrillion operations per second). IBM's current supercomputer, codenamed 'Roadrunner', has been built using components designed for the lowlier Sony Play Station 3.

2

A cone has only two surfaces.

2

When test-pilot Chuck Yeager first broke the 'sound barrier' in his rocket-powered Bell X-1 in 1947, he had to hide the fact that he had broken two ribs in a riding accident the day before.

2

There are only two months in recorded history that did not have a full moon – February 1865 and February 1999.

3

Feet – the maximum height to which an ancient Archimedean screw could lift water. This simple method of raising water, by a spiral screw contained in a tube, from a low level to a higher one is attributed to Archimedes in the 3rd century BC, but some believe that Nebuchadnezzar was the originator in the 7th century BC.

3.14159...

Pi – the ratio of the circumference of a circle to its diameter. While the value of pi has been computed to a trillion decimal places, normal scientific use normally requires no more than a dozen. Eleven decimal places are enough to calculate the earth's circumference to within a millimetre.

3.8

It's goodbye to the Moon! It is moving further away from Earth by about 1.5 inches (3.8cm) a year. This is caused by tidal effects, and consequently the Earth is slowing its rotation by about 0.002 of a second per day per century.

3.98

The degrees Centigrade at which water strangely starts to re-expand – having contracted as it has become colder.

4

The earth consists of four layers – the crust, the mantle, the outer core and the inner core. The lithosphere combines the crust, the earth's hard outer shell, and the top part of the mantle is between 3 and 40 miles (5 – 64 km) deep. Below the sea, the thickness is much smaller than that of mountain ranges. This is very thin compared to the overall depth of the mantle which is 1,800 miles (2,900 km), the outer core which 1,300 miles (2,200 km) and the inner core which is 800 miles (1,288 km). The core consists mostly of iron – the inner being solid, while the outer is molten at a temperature of 4,300 C (8,060 F).

4

There are four 'strokes' in an internal combustion engine – induction, compression, ignition, exhaust. This is called the 'Otto Cycle', named after the German scientist Nicolaus Otto in 1876.

4

Elements that Greek philosopher Aristotle (384-322 BC) thought made up our world – air, earth, fire and water.

5

There are five 'kingdoms' of biology – animals, plants, fungi, prokaryotes (bacteria and blue-green algae) and protoctista (algae, seaweeds, protozoa and mildews).

5

There are at least two eclipses of the sun in a year, and a maximum of five. A solar eclipse occurs when the Moon passes between the Earth and the Sun, so that the Sun is completely or partially blocked from view. Total eclipses happen when the Sun is completely obscured by the dark silhouette of the Moon. These are rare, and, as the sun disappears during the day, were treated with awe and suspicion in ancient times.

7

A rainbow is visible when the sun shines from behind the observer on to moisture in the atmosphere. The resulting arc spans the spectrum of seven colours: red, orange, yellow, green, blue, indigo and violet. A fully circular rainbow can only be seen from an aeroplane - with the plane's shadow in the centre.

8

Once light has been emitted from the Sun's surface, it takes about 8 minutes to reach Earth. For other planets in our solar system, it will take about 3 minutes to reach Mercury, and over 5 hours to get to Pluto.

8

The architecture of the early IBM computers, and subsequent microprocessors, were based on a storage system of 8 bits in a 'byte' (the word was coined by a Dr. Werner Buchholz in 1956 from the shortened pronunciation of 'by eight', and also because it was the smallest piece of information a computer could 'bite'). With the increased capacity of computer processing power, so too have the number of 'bits' in a 'byte' increased, along with attempts to rename the new units – 16 'bits' in a 'byte' can be called a 'plate', 18 'bits' a 'chomp', 32 'bits' a 'dinner' and 48 'bits' a 'gobble'; conversely, 2 bits are a 'crumb' and 4 'bits' a 'nibble'.

8.5

Minutes – the time it takes a Space Shuttle to go from standstill to 17,000 mph. The propulsion gasses leave the solid booster rocket at 6,000 mph – three times the speed of a high-powered rifle bullet.

10

The number of atomic particles in the universe is 10 raised to the power of 72, or, if you want to read it in full, 10,000,000,000,000, 000,000,000,000,000,000,000,000,000,000,000,000,000,000,000,00 0,000,000,000,000.

10

The top ten feet of the ocean hold as much heat as the entire atmosphere. However, at its deepest the ocean temperature is about 4 C (39 F) – just above freezing.

10.56

The speed record on the surface of the Moon is 10.56 miles per hour (17 kph). This was set by the Lunar Rover (moon buggy) on Apollo Space Missions 15, 16 and 17 which landed on the Moon between 1971 and 1972. On its last mission, the Lunar Rover was used for a period of about seven hours to retrieve 243.6 lbs (110.5 kg) of moon rock which was brought back to Earth.

12

There are twelve levels of wind strength as defined by Admiral Sir

Francis Beaufort in the eponymous scale he developed in 1805, that was not officially used until the famous voyage of the Beagle (1831 to 1836). Still in universal use today, the strongest is Hurricane Force 12 with wind speeds of 73-83 mph (117-134 kph); Force 11 - a violent storm with winds of 64-72 mph (103-116 kph), and the slightly less Force 10 winds of 55-63mph (86-102 kph) which would, on land, uproot trees and cause structural damage.

14

Carbon-14, the radioactive isotope of carbon, whose rate of decay can be measured so that the age of everything up to 40,000 years old can be accurately dated. Developed in America in 1949, it is universally known as 'radiocarbon dating'. The system of 'potassium-argon dating' is used for older materials, and is based on some of the radioactive potassium isotope decaying into the gas argon.

17

The number of space flights in America's Apollo programme, placing 30 astronauts in space and 12 on the moon, 11 of whom, interestingly, had once been Boy Scouts.

20

The time of twilight (which refers to dawn as well as dusk) varies considerably – depending upon the location and latitude of the observer. On the Equator it can go from bright daylight to night in only 20 minutes, while twilights can last several hours at the North and South Poles.

27.3

Days it takes the moon to complete a full orbit around the earth.

29

The number of years on Saturn that equates to an Earth year. This is because Saturn goes round the Sun very slowly, but is spinning very fast on its own axis. A Saturn *day*, conversely, is much shorter than that on Earth, at 10 hours and 14 minutes.

32

There are 32 calories in 100 ml of beer (both bitter and Guinness) while lager can be 40 calories. The equivalent calorie count for gin and vodka is 222, sherry is 136, white wine 74, champagne 76, and, leading the field, is Baileys Irish Cream at 350 calories.

32.2

Feet per second per second, the acceleration of a body falling due to the earth's gravity. However, under the influence of air resistance, a human body would reach its terminal velocity of about 120 mph after 14 seconds and 1,880 feet (573 m).

49.2

Percentage of the earth's crust that is oxygen. Oxygen also makes up 28% of our atmosphere.

50

LD-50 dose, the toxicological term that defines the level of drugs or poisons that kill 50% of victims.

72

Degrees °C (161.6 °F), the temperature at which food is heated for 16 seconds to achieve pasteurisation. UHT ('Ultra Heat Treated') milk, as well as other beverages, is heated at a temperature exceeding 132 °C (275 °F) for 1 or 2 seconds to achieve partial sterilisation, after which it then lasts for months.

87

If you stretch out flat a standard Slinky, the toy that walks down stairs by gravitational pull, it measures 87 feet (26.5 m) long. Originally invented in 1943 by US Naval engineer Richard James as a result of studying a torsion spring, he made 400 trial units of the toy which sold out in 90 minutes. A Slinky has other uses – such as in science education and it was also used a mobile radio antenna during the Vietnam War.

90

Ninety percent of all volcanic activity takes place in the world's oceans. The largest known concentration of active volcanoes is in the South Pacific in an area the size of New York State and which has 1,133 volcanic cones.

99

The oceans on Earth contain 99% of the living space on our planet, but only about 10% of this space has been explored. We know more about space than we do about our deep oceans. An estimated 80% of all life on earth is found under the ocean surface. 85% of the area and 90% of the volume are the dark and cold environment of the deep oceans.

148

To become a member of MENSA, a minimum score of 148 is required in the Cattell IIIB Intelligence Quotient (IQ) test, and membership places those who succeed in the top two per cent of the population. The maximum score is 161. MENSA has 25,000 members in the UK and Ireland, and 100,000 worldwide.

180

The temperature on the surface of Mercury goes through a dramatic range – by day it exceeds 430 C (806 F), but at night it plummets to minus 180 C (356 F).

193

Should you be able to drive to the Sun and keep to a speed of 55 miles per hour (88.5 kph), your journey would take about 193 years. It you wanted another unlikely 'space trip' and drove around one of Saturn's rings slightly faster at 75 miles (121 kph) per hour, this would take 258 days.

360

Degrees in a circle. The origins are thought to lie with the ancient Babylonians who worked on a sexagesimal numbering system using 60, rather than one based on 10. Also, 360 is readily divisible by every number between 1 and 10 – except 7. In order for a system to be divisible by all numbers including 7, there would have to be 2,520 degrees in a circle.

415

Feet (126 m) – the height of the tallest recorded Douglas fir (also know as the 'Oregon pine'). For various reasons, 19th century botanists found this tree difficult to classify, and the hyphen in its common name indicates it is not a true fir – not a member of the genus *Abies*. It was introduced to the UK in 1827 by famous plant hunter, David Douglas.

499

Although Cherrapunji, in east India, is credited with being the wettest place on earth with 499 inches (1,270 cm) of rain a year, it has problems with having enough potable water. The shortage is caused because it only rains for six months of the year, and when it does there is no reservoir, so the rain washes away down the hills, taking dirty topsoil with it.

1,000

The number of Earths that would fit into Jupiter's sphere – the biggest planet in our universe.

1,000

When our universe was created 13 billion years ago by the explosion of matter, known as 'Big Bang', the temperature at the centre, for a tiny part of a second just after the explosion, is estimated to have been 1,000 trillion degrees Celsius.

1,093

The number of US patents that American Thomas Edison (1847-1931) held on his inventions, ranging from phonographs, telephones, and electric railways to, most, significantly, electric lighting.

1,250

The highest wind speed on any planet is 1,250 mph (2,000 kph) on Neptune. On Saturn's equator winds can reach 1,118 mph (1,800 kph). In comparison, the strongest winds on earth are around 200 mph (322 kph).

1,250

The number of scanning lines on High Definition televisions, a great improvement on the 625 (UK) or 525 (US) lines being used at the moment. However, this increased definition has meant that many actors and actresses have rushed off to beauticians, and even cosmetic surgeons, in reaction to the harsh truths revealed. John Logey Baird's first part-mechanical television system in 1936 used only 240 lines.

1,300

Degrees Centigrade (2,372° F) – the temperature at which household bricks are baked.

1781

When the seventh planet from the Sun, Uranus, was discovered by the English astronomer Sir William Herschel in 1781, he briefly considered naming it 'George' *(Georgium Sidus)* in honour of his patron and England's King – George III. Amongst other proposed names were 'Herschel' after its discoverer. However, in 1851, its name was finally confirmed as Uranus – the Latinised version of the Greek god of the sky, and the father of Saturn (who in turn was father of Jupiter).

3,000

If you counted at the rate of one star a second, it would still take you three thousand years to count all the stars in the constellation.

3,000

The largest canyon system in the Solar System is Valles Marineris on Mars. It is more than 3,000 miles (4,830 km) long, and so would stretch from California to New York. In some places it reaches 3 miles in depth and is 200 miles (322 km) in width.

3,280

Feet (1,000 m) – the visibility distance below which mist is defined as fog.

4,708

The speed of sound in water is 4,708 feet per second (1,435 metres per second) - nearly five times faster than the speed of sound in air.

12,450

The average depth of the ocean is 12,450 feet (3,795 m), while the average height of the land is only 2,756 feet (840 m)

22,000

Miles – a geostationary satellite travels at approximately this altitude above the earth, and at a speed of 7,000 mph (11,000 kph).

35,813

In 1960 Jacques Picard and Don Walsh, in their manned bathyscape *Trieste*, reached the deepest point on the earth's surface – 35,813 feet below sea level, (10,916 m) in the Mariana Trench, in the north-west Pacific. The pressure on the hull was 8.62 tons per square inch (1.25 tons per square centimetre), or approximately 1,095 times that on the surface. Even at that incredible depth, small flounder like fish were seen by the crew.

36,000

Mph (58,000 kph) -- The speed of the spacecraft *Voyager 1*, which was launched by NASA on 5 September 1977 to undertake a grand tour of the outer planets. It is currently heading out into inter-stellar space and is travelling at approximately 36,000 mph, which makes it the fastest man-made object ever. However, at that speed it would take it 80,000 years to reach the nearest star, Alpha Centauri, which is 4.3 light years away.

99,787

Miles (160,657 km) – the circumference of the planet Uranus – compared with that of Earth which is 24,902 miles (40,092 km).

335,789

The Sun is 335,789 times the weight of Earth, while Jupiter is 319 times heavier. The Earth's mass is 13,169,619,000,000,000,000,000,000 lbs (5,973,700,000,000,000,000,000,000 kg).

360,000

Thermal megawatts, the power surge up from 200 megawatts that blew the 1000-ton steel top off the reactor at Chernobyl in 1986, in the world's worst nuclear accident.

400,000

Hydrogen bombs (or 1,500,000 power stations working for 8 months) that would need to explode to have the same heating effect on the world as *El Niño*, the warm water current off South America that is half as large again as the United States and which, occuring only every few years, has a dramatic influence on weather patterns. *El Niño* is Spanish for 'the child', and is so named after the baby Jesus because the effect tends to start around Christmas time.

660,000

The speed in mph (1,062,600 kph) at which Earth is travelling through space.

2,200,000

Light years, the furthest distance away at which stars are visible to the naked eye. Only 2,500 stars can be seen at any one time.

3,200,000

Miles (5.1 million km) – the closest distance that Halley's Comet has ever passed to earth was in 837 AD. It last appeared in 1986, and will next be visible in 2061. It takes Halley's Comet 76.1 years to orbit the sun.

2,304,000

The deepest part of any of the world's oceans is seven miles – the Mariana Trench in the Pacific – where the pressure is more than 2,304,000 pounds per square foot (11,250,000 kilos per square metre), or the equivalent of one person trying to support 50 jumbo jets

14,000,000

The temperature at the core of the 4.5 billion year old Sun is 14 million degrees C (22.5 million F). Its luminosity is 390 billion billion megawatts.

5,600,000

Parts on the *Apollo 8* space rocket, which also had one and a half million systems, subsystems and assemblies. Even if all functioned with 99.9 per cent reliability, a worrying fifty-six hundred defects could be expected!

32,000,000

The times a car tyre rotates in its average lifetime.

300,000,000

Between 300 and 500 million people a year worldwide are affected by malaria, whichs sadly causes about 2 million deaths – 90% of which occur in Africa and 6% in Afghanistan, Brazil, Colombia, Sri Lanka and Vietnam. 41% of the world's population are at risk of acquiring malaria.

366,000,000

Miles (590,000,000 km) that all US Space Shuttles have travelled cumulatively – spending a total of more than three years in flight, and carrying 600 passengers and pilots.

5,000,000,000

The odds of a human being killed by space debris falling on Earth is 1 in 5 billion, and the risk of being struck by a falling meteorite is once every 9,300 years.

983,571,056.4

Feet per second (299,792,485 metres per second) or one foot per nano second – the speed of light. Thus, light travels 5,878,625,373,183.61 miles (9,460,730,472, 580.8 km) in a year.

9,192,631,770

The beats per second of a cesium atom in an atomic clock – the world's most accurate gauge of time, and the standard used by broadcasters and satellite signals.

3,700,000,000,000

The highest manmade temperature achieved so far is by the 'Z' machine at Sandia National Laboratories in New Mexico – created to test materials in extremes – and which has achieved temperatures in the region of 3.7 billion degrees F. The 'Z' machine can release 80 times the world's electrical energy in a few billionths of a second, but is comparatively economical in so doing – using only the total energy consumed by 100 US households over two minutes.

24,792,500,000,000

Miles (39,916,000,000,000 km) – the distance to the nearest star from earth.

...And, finally,

if we accept infinity as a concept, then, along with endless other permutations, it must be that somewhere else in the universe there is somebody just like you, reading exactly the same book which has been written by very much the same authors!

If you've enjoyed this book, why not become a member of the 'Numeroids' community by signing up for free to our dedicated website?

www.numeroids.com

Where you can

- Submit your own interesting 'Numeroid'
- Look other 'Numeroid' contributions
- Enjoy 'Numeroid' competitions
- Receive regular 'Numeroid' updates